The
Milkweed
Lands

An Epic Story of One Plant
Its Nature and Ecology

By Eric Lee-Mäder

Illustrated by Beverly Duncan

Storey Publishing

The mission of Storey Publishing is to serve our customers by publishing practical information that encourages personal independence in harmony with the environment.

Edited by Deborah Burns
Art direction and book design by Jessica Armstrong
Text production by Jennifer Jepson Smith
Illustrations by © Beverly Duncan

Text © 2023 by Eric Lee-Mäder

© Storey Publishing, LLC/Foreword by Joan Edwards

Storey books are available at special discounts when purchased in bulk for premiums and sales promotions as well as for fund-raising or educational use. Special editions or book excerpts can also be created to specification. For details, please send an email to special.markets @hbgusa.com.

Storey Publishing
210 MASS MoCA Way
North Adams, MA 01247
storey.com

Storey Publishing, LLC is an imprint of Workman Publishing Co., Inc., a subsidiary of Hachette Book Group, Inc., 1290 Avenue of the Americas, New York, NY 10104

ISBNs: 978-1-63586-436-6 (hardcover); 978-1-63586-437-3 (ebook)

Printed in China through World Print
10 9 8 7 6 5 4 3 2 1

Library of Congress Cataloging-in-Publication Data on file

CONTENTS

Foreword

With their milky, toxin-laden sap, feathered windborne seeds, and one of the most complex flowers in the plant world, milkweeds are fascinating in and of themselves. But Eric Lee-Mäder takes the plant into a new realm in *The Milkweed Lands*. He presents a rich tapestry of milkweed's life events against the backdrop of changing seasons and in multiple settings, from seemingly insignificant roadside ditches to the flood plains of the Mississippi to restored habitat in California's Central Valley.

Milkweeds are never far away. At least one of the many species of *Asclepias* grows in almost every state and should be known by all, but few of us recognize the depth of interdependency brokered by these plants. Best known as the only food of monarch caterpillars, milkweeds provide a lifeline for that iconic endangered butterfly, but, as *The Milkweed Lands* demonstrates, the interactions go beyond this one relationship.

Milkweeds support a rich array of life both above- and belowground. Multilevel connections among herbivores, predators, competitors, and pollinators play out over and over in the milkweed world and exemplify how one species links to hundreds of others. Weaving together stories and colorful illustrations, *The Milkweed Lands* vividly illustrates the interconnectedness of nature.

Careful observation yields pleasure in discovery and an understanding of the diversity of interdependent life. Although this book focuses on the network of one plant, its message of connectivity applies broadly to all species, so critical amid our current global decline in biodiversity. *The Milkweed Lands* exemplifies the importance of knowing nature in order to preserve it.

JOAN EDWARDS, PhD
Samuel Fessenden Clarke Professor of Biology
Williams College

The Milkweed Lands

The milkweed is a displaced citizen in its own land. Where once it owned the continent, it's now a kind of vagrant, occupying the botanical equivalent of homeless encampments.

Milkweed lives out along distant stretches of railroad tracks, far from people and mowers, among overgrown grasses and the odd bushy sweet clover. It endures around the empty, buckled parking lots of abandoned Midwestern shopping malls, where teenagers spray-paint the walls at night and bits of plastic rubbish slowly decompose into a new kind of soil.

Growing up, I haunted these same kinds of places. And as an adult I find myself increasingly returning to the magical, unpeopled haunts of my youth, looking for milkweeds in their messy natural state, alongside other fugitive plant and animal friends.

This book is a dispatch from those Milkweed Lands, or perhaps more appropriately a retelling of their grand, feral existence. Illustrator Beverly Duncan and I have attempted a version of ancient storytelling (in this digital age), with me speaking dramatically and waving my hands around the fire while she adorns the cave walls with epic paintings of milkweeds living out their heroic lives.

As you read this, we hope you take some inspiration from the fact that so many wild things are still persisting in these jumbled-up circumstances, evolving, every day, all the time. As we all are. Life is fragile, miraculous, and resilient. It's more than the sum of its parts; more than any one thing alone.

Just ask the milkweed.

—ERIC LEE-MÄDER

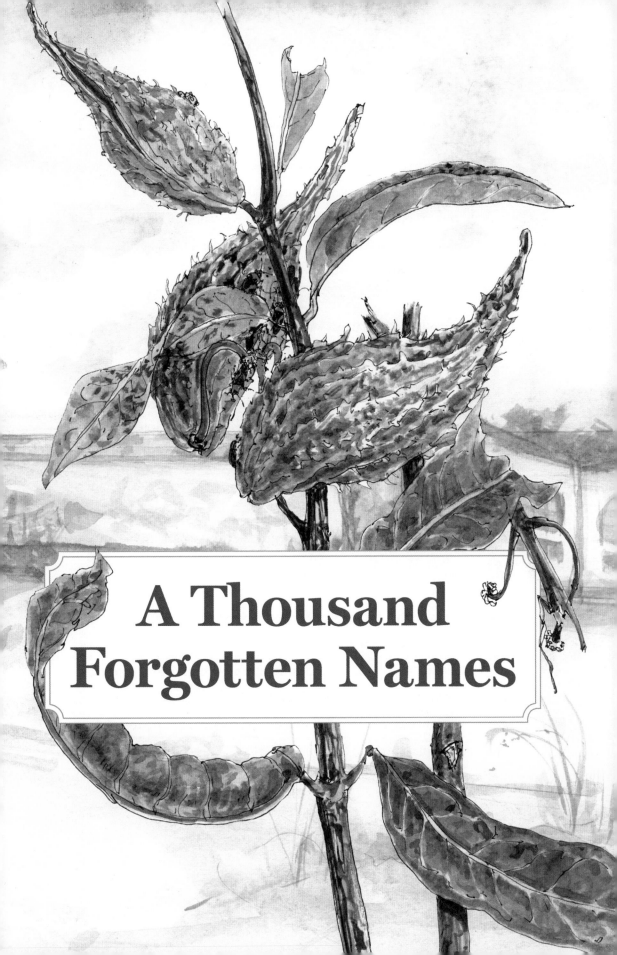

A Thousand Forgotten Names

THE FIRST PEOPLE who lived with North America's native milk-weeds gave them their earliest names—*o-wah-kwen'stah, waxca-xca, pannun'pala,* along with a thousand other appellations. Like the tales of those people, the story of milkweed was appropriated by those who came later, the author of this book being no exception.

The commandeering of the milkweed story becomes especially obvious when you consider that the scientific naming system for these American plants is entirely, and almost haphazardly, a product of European self-regard (with a little Middle Eastern geography thrown in).

Louis Hébert, a French farmer who settled on the Saint Lawrence River, near what is now Quebec City, and who is sometimes claimed to be Canada's first pharmacist, may have been the original conduit for intro-ducing American milkweeds to Europe. Hoping to determine the plant's medical potential, Hébert sent seeds to Paris, where doctor and botanist Jacques-Philippe Cornut grew and studied them. Cornut went on to pub-lish the first written account of American milkweeds in his 1635 treatise, *Canadensium plantarum aliarúmque nondum editarum historia.*

Common milkweed, the species Cornut most likely first encoun-tered, did indeed have a pharmacological tradition among Indigenous people of North America and early Western settlers. Records refer to its use as a treatment for coughs and congestion, rheumatism and pneu-monia, and a range of other ailments. Based on these notions, a century later Carl Linnaeus, the Swedish inventor of modern nature taxonomy, is thought to have imagined upon *Asclepias* as a proper genus name for these plants—in homage to the Greek hero-god of medicine, Asclepius, most often remembered for his snake-entwined staff, a symbol still associated with Western medicine today.

To complete his official two-part scientific name for common milkweed, Linnaeus is also thought to have added the epithet (species name) *syriaca*, referring to Syria, for the plant's resemblance to a Middle Eastern plant also being cultivated in Europe at that time (the term possibly reflecting confusion over the plant's origin). Thus North America's most abundant, widespread, and well-known milkweed came to be named for a Greek god and a Middle Eastern nation.

Much has been written of the Eurocentric bias of Linnaeus. Unimpeded by this—or by the odd-fitting scientific name—the humble and handsome common milkweed, with its multitude of lost titles, eventually escaped and spread across Europe. It now persists in feral populations alongside highways and industrial parks. It is even said to grow lavishly along the Dnieper River, downstream from the Chernobyl radioactive exclusion zone.

Long severed from ancestors, perhaps in time emigrant milkweeds will become new kinds of plants, a future branch on an evolutionary tree, something new to be named.

The Dogbane Family

Along with three hundred other genera (and more than five hundred species), the *Asclepias* genus is part of the larger dogbane family, Apocynaceae, whose members occur on every continent except Antarctica. Many in this broad group produce toxic latex sap and are otherworldly, leafless succulents. Some grow low to the ground, others as lanky, sprawling shrubs.

Dogbanes are frequently, but not always, desert plants. They all possess five-petaled flowers, joined into a tube at the lower end, with each flower hugged by a gamosepalous base—a sometimes elaborate lobed covering at the point where flower meets stem. As a whole, it's a potent family: a source of medicines and poisons, hallucinatory drugs, food, fiber, and rubber. Various family members have been used to make everything from poison-tipped arrows to experimental cancer treatments.

Striking Similarities

Although geographically dispersed, members of the dogbane family are sometimes strikingly similar in their resemblance. I experienced an uncanny recognition one morning in southern India, on a work assignment at a bee research facility, when I noticed an enormous milkweed, taller than I was, growing next to the road.

Despite belonging to a completely different genus, the plant, known as crown flower (*Calotropis gigantea*), resembles a superhero version of North America's common milkweed. Growing up to 13 feet (4 m) in height, crown flower has leaves that are broader than those of North American milkweed, and its flowers are bigger, showier, and even more fragrant. It's a cosmopolitan plant, occurring across a vast range in tropical Asia and Africa. It is also an exotic weed in other tropical places across the globe, including Hawaii. There it and the also nonnative monarch butterfly have formed a kind of blended immigrant relationship with one another.

MILKWEED'S FAMILY
Apocynaceae (the Dogbanes)

There are about 5,556 species in this large family, including all the North American milkweeds.

The GAMOSEPALOUS base, with fused sepals

REPRESENTATIVE FLOWERS

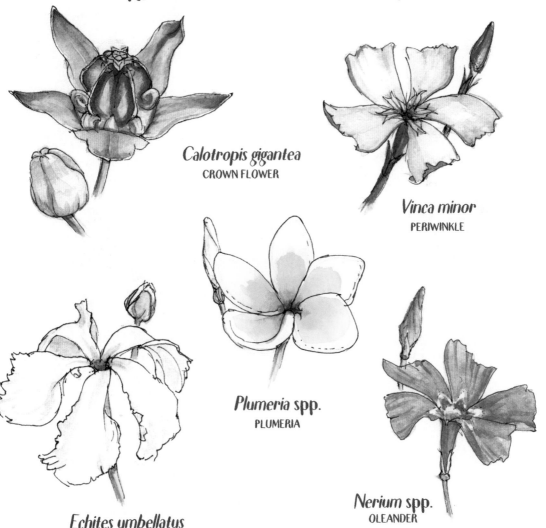

Calotropis gigantea
CROWN FLOWER

Vinca minor
PERIWINKLE

Plumeria spp.
PLUMERIA

Echites umbellatus
DEVIL'S POTATO

Nerium spp.
OLEANDER

NORTH AMERICAN MILKWEEDS
REPRESENTATIVE SPECIES

Asclepias viridis
GREEN MILKWEED

Asclepias erosa
DESERT MILKWEED

Asclepias purpurascens
PURPLE MILKWEED

Asclepias variegata
WHITE OR REDRING
MILKWEED

Asclepias speciosa
SHOWY MILKWEED

Asclepias asperula
ANTELOPE-HORNS MILKWEED

Asclepias californica
CALIFORNIA MILKWEED

Asclepias fascicularis
MEXICAN WHORLED OR
NARROWLEAF MILKWEED

Asclepias tuberosa
BUTTERFLY MILKWEED OR
BUTTERFLY WEED

Asclepias syriaca
COMMON MILKWEED

Asclepias lanuginosa
WOOLLY MILKWEED

This species is unusual,
with no horns or hoods
below the stigmatic disk.

Toxin or Tonic?

Our North American milkweeds in the *Asclepias* genus (of which there are more than 90 recognized species) are arguably most famous for two things: first, for their role as food for the caterpillars of monarch butterflies (*Danaus plexippus*), and second, for their production of cardenolides, potent steroids that can disrupt the life functions of vertebrate animals unlucky enough to feed on them. These two attributes are intrinsically linked.

Many plants produce toxic or ill-tasting chemicals to discourage herbivores. Milkweeds take this strategy especially seriously. Rangeland horses in the West feeding on whorled milkweed (*Asclepias subverticillata*) may lose muscle control and experience violent spasms. A horse's heart may slow, and it may go into respiratory paralysis. However, it seems that most livestock are immediately put off by the bitter taste and will refrain from eating the plant unless starving.

In some circumstances, milkweeds don't need to be ingested to unleash their chemical defenses. Accidentally rubbing your eyes after handling tropical milkweed (*Asclepias curassavica*), a South American species, may cause painful eye lesions, eye swelling, blurred vision, sensitivity to light, and, possibly, temporary blindness. Not only do milkweeds produce these powerful chemical defenses, but they also discharge thick, gluelike latex sap that gums up the mouths of plant feeders and sticks to clothes and skin.

Yet a lot of animals decidedly do eat milkweeds, the most famous, of course, being monarchs. Animals that feed on milkweed tend to advertise their dangerous diet in the loudest visual ways possible, via a mechanism called aposematism.

Dangerous Colors

Aposematism is the technical term for the bright coloration that prey animals use to inform a potential predator that they are toxic, dangerous, or otherwise unpalatable. This trait is everywhere in the animal world once you start looking for it. Skunks walk around adorned in racing stripes as a kind of signal that they are too important to be messed with, as do colorful red, black, and yellow coral snakes.

So too, various insects display their ability to eat toxic fare, such as milkweeds, without harm, and to absorb the plants' chemical defenses as their own. Hence, monarch butterflies have their striking orange-and-black wings, and the daftly colored calotropis or painted grasshopper (*Poekilocerus pictus*) is a large, fluorescent-striped marvel of orange, blue, and chartreuse that feeds on crown flower in India.

Milkweed on the Menu

People, too, sometimes eat milkweed, especially common milkweed, which is lower in cardenolides than some of its relatives. Preparation usually includes repeated blanching in water to reduce the prospect of becoming ill. Once just a subsistence food source, milkweed is being rediscovered by new generations of adventurous foragers and chefs who cook young shoots as if they were asparagus or serve up boiled green pods as a novel vegetable.

WARNING COLORATION

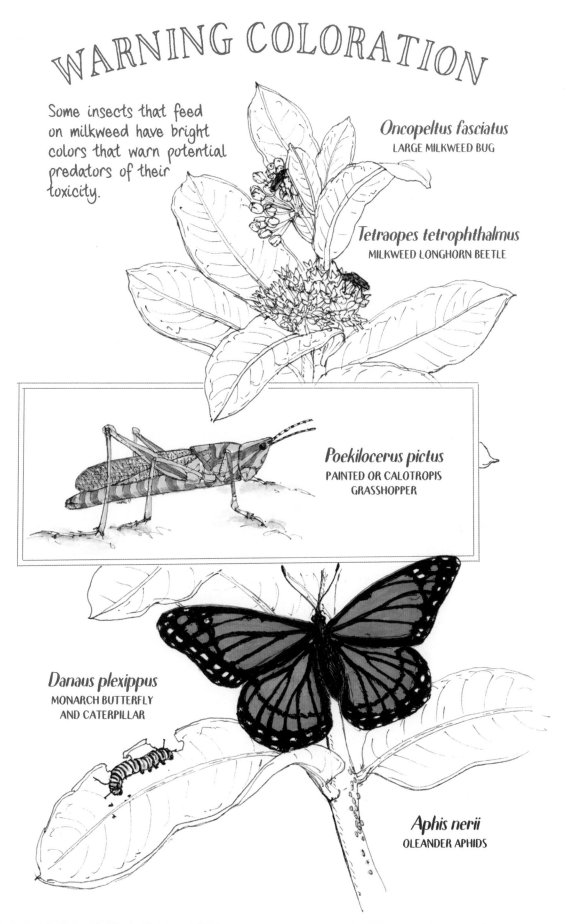

Some insects that feed on milkweed have bright colors that warn potential predators of their toxicity.

Oncopeltus fasciatus
LARGE MILKWEED BUG

Tetraopes tetrophthalmus
MILKWEED LONGHORN BEETLE

Poekilocerus pictus
PAINTED OR CALOTROPIS GRASSHOPPER

Danaus plexippus
MONARCH BUTTERFLY
AND CATERPILLAR

Aphis nerii
OLEANDER APHIDS

CARVER'S MILKWEED

George Washington Carver advocated eating wild plants—including milkweed—during World War II.

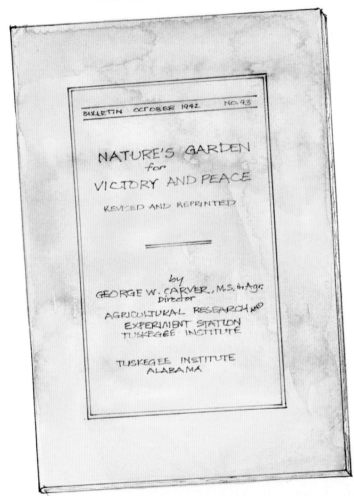

BULLETIN OCTOBER 1942 No. 43

NATURE'S GARDEN
for
VICTORY AND PEACE

REVISED AND REPRINTED

by
GEORGE W. CARVER, M.S. in Agr.
Director
AGRICULTURAL RESEARCH and
EXPERIMENT STATION
TUSKEGEE INSTITUTE

TUSKEGEE INSTITUTE
ALABAMA

Carver's Milkweed Tips:

MILK WEED FAMILY
(Asclepiadaceae)

"Have always held a high place as a delicious food; cut just before the leaves are half grown, prepare like asparagus tips. They improve all mixed greens. They are also choice boiled or steamed until tender and served with mayonnaise or French dressing; and they are equally fine in any mixed salad.

"They are good also in a puree of vegetables, bouillon cubes or gelatinized vegetables."

CARVER'S OTHER VICTORY WEEDS
Excerpted from Carver's Guide

1. *Plantago major* (dooryard plantain)
2. *Phytolacca decandra* (pokeweed)
3. *Rumex crispus* (curled dock)
4. *Antennaria plantaginifolia* (rabbit tobacco)
5. *Elephantopus tomentosus* (giant thistle)
6. *Chrysanthemum leucanthemum* (ox-eye daisy)
7. *Matthiola incana* (stock)
8. *Oenothera biennis* (evening primrose)

Victory Milkweed

The tradition of eating milkweed experienced a brief revival in 1942, when famed Tuskegee University scientist George Washington Carver published a pamphlet on how to prepare edible weeds. The purpose of the pamphlet was to encourage civilians to forage for wild plants ("victory weeds") so that cultivated vegetables could be given priority for the overseas fighting forces. There is even a relic photo that appears occasionally, showing Carver and Henry Ford preparing to eat a meal of milkweed and other wild plants together. Ford was fascinated by many of Carver's innovations, including a process for manufacturing rubber from milkweed latex.

Although most famously (and mistakenly) credited for inventing peanut butter, Carver was foremost a conservationist devoted to helping poor farmers protect and enhance their soils. He published a guide to nature education for primary school students that still compares favorably to any grade-school curriculum produced today, and he contributed enormously to agricultural innovations that helped win the war. Born into slavery, he had been kidnapped and ransomed as a child and was thought to have been severely abused as an infant by his enslaver. Despite unbelievable odds, he rose through white academic ranks and became one of the finest minds the United States ever produced.

Common milkweed (*Asclepias syriaca*) might better be named Carver's milkweed (*Asclepias carverii*).

Changing Fortunes

The fortunes of milkweed have been a series of boom-and-bust cycles. Whereas earlier peoples saw these plants' healing power, the past century has witnessed a determined effort to eradicate them.

Examples of this are everywhere, as in the beautifully illustrated 1963 hardcover book *Pasture and Range Plants*, published, oddly, by the Phillips Petroleum Company in Oklahoma during an era when oil and cattle went hand in hand. It says of the gorgeous, minimally toxic, native butterfly milkweed (*Asclepias tuberosa*):

> *It is an invader on our native ranges. Chemical sprays and sound management will provide adequate control.*

Similarly, five milkweed species are featured in recent editions of *Weeds of the West*, a thick picture ID book published by the Western Society of Weed Science. The book, long considered a definitive authority on what constitutes a rangeland weed, has included the tall and beautiful swamp milkweed (*Asclepias incarnata*), which, as the name suggests, grows in natural wetlands where cattle would not be likely to graze.

Even though monarch butterfly populations began a tremendous decline over the past 20 years, milkweed continues to be targeted for elimination. As recently as 2020, a Midwest farming television series, *Ag PhD*, produced by a pair of South Dakota brothers, recommends the herbicide glyphosate to get rid of milkweed in corn and soybean fields.

Beyond agricultural lands, milkweeds have long been restricted under local municipal weed ordinances in numerous Midwestern US cities and suburbs. In places such as Evanston, Illinois, city weed

WHAT IS A WEED?

In modern English, a weed is a plant that is considered undesirable in a particular area. The origins of the word itself, however, suggest something altogether different.

To speak of *weide* in old Dutch or German, from which the word arose, is to speak of meadows themselves, places of pastures and wildflowers, places of life.

ordinances established in the 1950s specifically mentioned milkweeds and threatened would-be prairie gardeners with having their yards forcibly mowed by city staff and being sent a bill for the "service."

"Best Honey Yielding Plant"

The desire to get rid of milkweeds marks a break with an earlier era in which the plants served obvious and valuable utilitarian functions. Among these was honey production. Writing of *Asclepias tuberosa* in an 1887 issue of *American Apiculturist* magazine, beekeeper James Heddon opined:

> *If there is any plant to the growing of which good land may be devoted exclusively to honey production, I think it is this. I would rather have one acre of it than three of sweet clover. It blossoms through July and the first half of August, and the bees never desert pleurisy root* [Asclepias tuberosa] *for basswood or anything else. The blossoms always look bright and fresh, and yield honey continuously in wet and dry weather. Bees work on it in the rain, and during the excessive drought of the past season it did not cease to secrete nectar in abundance. It is the best honey yielding plant with which I am acquainted, white clover and basswood not excepted.*

Praise of milkweed honey continued into the next century in 1940s editions of *The ABC and XYZ of Bee Culture*, a thick encyclopedia of beekeeping published by the A. I. Root Beekeeping Supply Company. The author remarked of common milkweed: "The honey is excellent and compares well with that obtained from raspberry. It is white, or tinged with yellow, and has a pleasant fruity flavor somewhat suggestive of quince, with a light tang."

Keeping the Navy Afloat

For a brief moment in time, milkweed was highly sought after for more than honey. With the entry of the United States into World War II, military equipment manufacturers faced a two-part problem: a sudden need to produce thousands of naval life vests, and a blockade of Southeast Asia, the traditional source of buoyant kapok tree fiber (the standard life-vest filling of that era).

The solution turned out to be milkweed seedpod fiber. To fill the need, the Milkweed Floss Corporation of America was quickly established in Petoskey, Michigan, a small working-class town at the far northern end of the state on Lake Michigan. For a few years local community groups gathered for picnics and milkweed-floss harvests. Children from across the Great Lakes region were mobilized with onion sacks to pick ripe seedpods for 15 cents per bag. Newspapers featured ads with headlines such as "Pick Milkweed Pods for Victory—for Profit."

New machines, milkweed gins, were invented to remove seeds and debris mechanically from the floss, while train cars and overloaded trucks delivered more than 12 million pounds of dried seedpods to the processing plant. For lack of storage alternatives, sacks of pods were lined up in the nearby county fairgrounds and hung by the hundreds on roadside fences in the town.

Today in Petoskey the old, unpretentious concrete building that housed this activity still stands, long since sold off and repurposed for some other industrial use. Directly across the street, next to a vacant barn and a desolate stretch of railroad tracks, is an overgrown patch of tall fescue, goldenrod, and common milkweed.

POD AND FLOSS

During World War II, milkweed was harvested and used in life vests.

NOW BUYING
RIPE MILKWEED PODS
15¢ Per Sack
SPECIAL OPEN MESH 50 LB. ONION SACKS
FILLED TO MAXIMUM CAPACITY

Wm Metzger   Stanley Cherry   Vern Welch
Ira House   Darrell Fleming   Claude Wilson

PICK PODS ● BUY WAR BONDS
War Hemp Industries – Petoskey, Michigan

Picking dried pods in early autumn

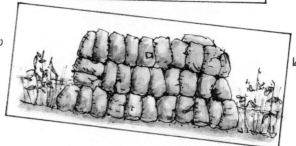

Bags of pods stored to dry before transport to the factory

The old factory in Petoskey, Michigan, still stands.

During World War II, milkweed life vests saved lives.

More recently there's been a small boutique industry for milkweed floss, both in Canada and the United States, where it is crafted into marvelous hypoallergenic pillows and comforters. One company that specializes in this even markets its bedding specifically for private jets and yachts: a kind of status artifact that is completely foreign to my socioeconomic sensibilities, to the child of poverty I once was.

The Herbicide Era

A half century on, I've managed to stumble with dumb luck into a better life for myself. Unfortunately, the milkweeds haven't. Researchers John Pleasants (University of Iowa) and Karen Oberhauser (University of Wisconsin) estimate that there has been an almost 60 percent decline in milkweed abundance in the Midwest since 1999 due to increases in agricultural herbicide use.

In a prechemical era, milkweeds managed to grow tenaciously in and around crop fields. Though periodically plowed up by cultivation equipment, some of those plants always resprouted from tough root systems. With the rise of corn and soy genetically modified for herbicide tolerance, crop fields have come to be plowed less and sprayed more. The milkweeds now struggle.

More poignantly, the estimate by Pleasants and Oberhauser leaves out the earlier, continent-wide purge of milkweeds that occurred when the original prairie sod was first plowed up. The true reckoning of how much milkweed we've lost in the past two hundred years must be staggering.

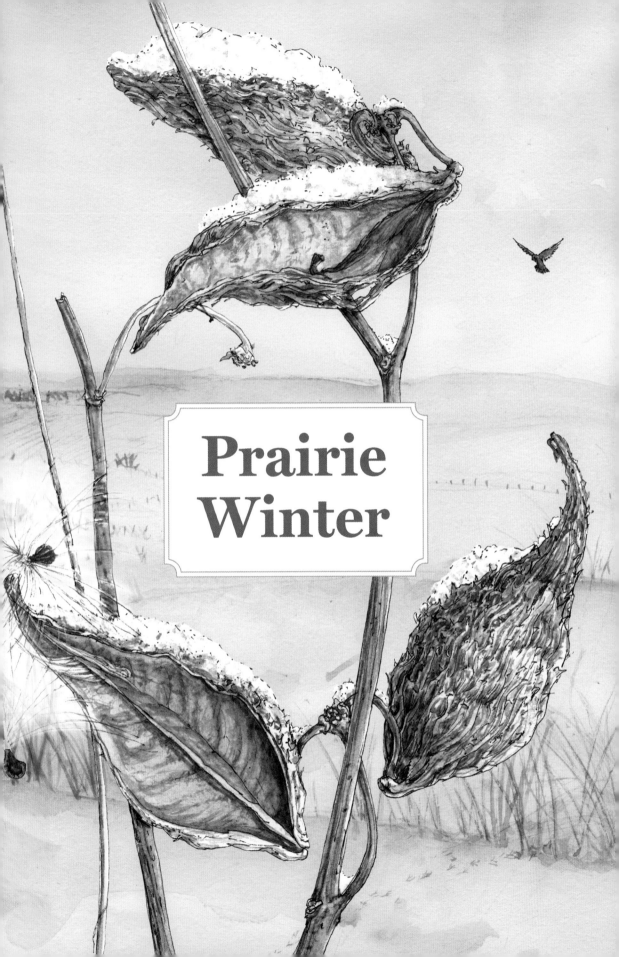

Prairie Winter

THE NORTHERN PRAIRIE WINTERS of my childhood were fero-
cious. Old Polaroids of blizzards, and a lifelong terror of river ice
imprinted on me by falling through it, offer confirmation of how harsh
the season could be.

Prairie winters are also revelatory. The snow records the comings
and goings of countless birds and small mammals that in the warmer
months are camouflaged by tall grasses. A lot of these tracks look like
miniature snowshoes.

The common white-footed mouse (*Peromyscus leucopus*) travels by
night on snow surfaces. Weighing less than an ounce, it is sure-footed
and nimble and likes to climb things.

A Blanket for a Mouse

These are most industrious mice. I once found a winter nest near a
cornfield, in the low crook of a shelterbelt tree. The base of the structure was a disused bird nest, perhaps built by a brown thrasher. The
bird is common on the northern plains, breeding in summer and feeding on grasshoppers and big ground beetles. Here, probably in early
February, the nest had been retrofitted with a thick dome roof of milkweed seed floss, compacted into a kind of blanket, with the edges neatly
tucked in.

For some reason, white-footed mice winter in these kinds of nests,
even though it would seem easier just to hunker down belowground.
Years after finding it, I still try to imagine life inside that nest through
howling winds and bitter cold. Such a tiny creature, with a heart
smaller than a pea, slumbering.

Milkweed floss

WINTER SURVIVAL SKILLS
OF THE WHITE-FOOTED MOUSE

Primarily nocturnal, this mouse is an excellent tree climber. It prefers to inhabit mixed stands of brushy deciduous trees. There it can find aerial nesting habitat and a variety of foods: edible seeds, nuts, dried berries, insects, and even fungi. It may store these items in a cache near the nest site.

In addition to plant fibers, such as milkweed floss, this mouse may build nests of grass, leaves, moss, bark, and even animal hair.

Adult mice sometimes perform high-pitched vocalizations and drum their front feet to warn other mice of danger.

The white-footed mouse is exceptional for its moonlit snowshoe comings and goings. Most other rodents, such as meadow voles, travel beneath the snow. Like the white-footed mouse, they may use a fluffy mix of fine grasses and milkweed floss to insulate their nests.

Hidden Highways

Larger than mice but smaller than rabbits, meadow voles are mostly either ignored or maligned. Still, they lead interesting, secretive lives, traveling along narrow, grass-covered pathways worn into the thatch by generations of their kind. Unless you crawl on hands and knees, spreading apart the grass as you go, you will rarely see these vole roads.

Even after deep snow covers the prairie, voles stick to these routes, as if traveling along crystal corridors. Foxes and coyotes can detect and pounce upon prey beneath the snow, knowing the vole roads the way humans know the streets of their town. Raptors are believed to locate them partly by the ultraviolet light reflection of vole urine.

Voles also have a curious habit of creating winter food caches, messy hoards of bulbs and thick rhizomes (horizontal underground stems), along these paths. I discovered this on my own farm in Washington State, when nearly a hundred showy milkweed (*Asclepias speciosa*) rhizomes that I planted went missing one winter. Walking in the tall meadow, I found a food cache that included a few of my rhizomes, along with other wild roots and seeds.

The cache belonged to a Townsend's vole (*Microtus townsendii*). That a mammal would feed on milkweed is surprising, given the plants' potent chemical defenses, but it is not the only rodent to do this.

UNDER THE SNOW COVER
A MAP OF PRAIRIE VOLE PATHWAYS

Milkweed rhizomes are one winter
food source for prairie voles.

A commuter among thatch tunnels

A nest of grasses

A food cache

CONTENTS OF A CACHE

Cirsium discolor
(field thistle) seeds

Carex spp.
(sedge) seeds

Psoralea esculenta
(prairie turnip)
roots

Asclepias speciosa
(showy milkweed)
rhizomes

Helianthus maximiliani
(Maximilian sunflower) seeds

A PRAIRIE WINTER COMMUNITY
AROUND THE ROOTS OF MILKWEED

ALGAE
0.002–0.025 mm

QUEEN BUMBLE BEE
0.8–1.3 inches
20.0–33.0 mm

ANTS
0.08–0.6 inch
2.0–15.0 mm

EARTHWORMS
0.4–15.0 inches
1.0–38.0 cm

NEMATODES
0.001–0.4 inch
0.3–10.0 mm

FLY LARVAE
0.6–1.4 inches
15.5–35.5 mm

SYMPHYLANS
0.08–0.4 inch
2–10 mm

ISOPODS
0.04–1.2 inches
1.0–30.0 mm

PROTURANS
0.02–0.06 inch
0.6–1.5 mm

THRIPS
0.02–0.08 inch
0.5–2.0 mm

MOLE CRICKETS
1.0–2.0 inches
25.0–50.0 mm

DIPLURANS
0.08–0.3 inch
2.0–7.0 mm

BEETLES
0.01–5 inches
0.25–127 mm

BACTERIA
0.00005–0.002 mm

FUNGAL HYPHAE
4–6 microns
0.004–0.006 mm

CENTIPEDES
0.12–12 inches
3.0–30.0 cm

BRISTLETAILS
0.4–0.5 inch
10.0–12.7 mm

PROTISTS
0.0002–0.006 inch
0.005–0.16 mm

ROTIFERS
0.004–0.02 inch
0.1–0.5 mm

POT WORMS
0.2–0.6 inch
5.0–15.0 mm

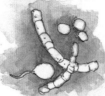

A Civilization Beneath the Frost Line

Also buried beneath the snow is a basic fact of winter: Even as the ground freezes from above, it is constantly thawing from below, due to our planet's inner heat. Squeezed between these opposing forces is the majority of life on Earth.

Indeed, ecologists have observed that one cubic foot (0.03 cubic meter) of temperate prairie or meadow soil contains more living things than does the entire aboveground Amazon rainforest. One cubic yard (0.76 cubic meter) of soil may contain thousands of individual fly larvae during much of the year, including march flies, soldier flies, crane flies, soil midges, picture-wing flies, and more. More than a billion bacterial cells may be present in a single gram of soil, especially near plant roots.

Fungal hyphae (strands or filaments of fungal cells) are possibly the largest and oldest continuously living creatures on Earth. Some connected networks cover dozens or even hundreds of acres. And protists—neither plant nor animal nor bacteria—are their own kingdom of some sixty-five thousand known species of various shapes, shells (or not), and forms of locomotion.

On a larger scale, male mole crickets construct underground auditoriums with complex acoustics in which to perform their courtship symphonies. Moles, meanwhile—beautiful, fragile, and maligned— have incredible lungs to extract oxygen in subterranean galleries, an almost supernatural sense for ground vibrations, velvetlike fur that remains spotless even in muddy conditions, and brief, mysterious lives.

Winter Denizens

Even in midwinter, this life goes on. A few exceptional invertebrate soil dwellers are active and conspicuous enough to be seen on warm days with a bit of soil melt. Various springtails, for example, tiny six-legged arthropods, may emerge from leaf litter in the soil to crawl on the surface of snow.

Springtails are named for their tail-like furcula, a body part that can quickly catapult them (in what appears to be an uncontrolled trajectory) away from danger. Their tiny size, the dark coloration of some species, and their jumping action gave rise to their other common name, snow fleas.

A secret to winter survival for snow fleas is their production of the amino acid glycine, which functions as a sort of antifreeze. Additionally, while springtails come in a wide variety of colors, depending on the species, those that are most highly active in winter, the snow fleas (especially those in the genus *Hypogastrura*), are blue-black, enabling their 2- to 3-millimeter-long bodies to absorb heat from the sun.

On the prairie, the sight of snow fleas is a harbinger of longer days. Late February in particular produces the kind of sunny, blustery days and occasional melts that reveal more soil life. It's a time when small milkweed bugs (*Lygaeus kalmii*) can sometimes be found deep at the base of hummock-forming grasses such as prairie dropseed (*Sporobolus heterolepis*), or even in the dried empty husks of broken milkweed seedpods beneath the snow. In winter, they enter a kind of stationary torpor.

One might wonder if they are dreaming of summer.

LIFE IN THE THATCH LAYER

Milkweed bugs and other tiny soil dwellers have different strategies for enduring a deep freeze.

SPRINGTAILS

Some springtails are covered with fine hairlike filaments or with scales like those on butterfly wings.

COLLEMBOLA: THE SPRINGTAILS

These tiny arthropods use antifreeze-like proteins in their bodies to survive the winter cold. They are beneficial creatures, feeding and processing pollen and decaying matter and bacteria in the soil.

Some COLORS of springtails:

Springtails that live on the surface of the soil are usually dark in color, while those that live deeper down tend to be pale.

HOW THE FURCULA WORKS

The furcula (Latin for "little fork") is the springlike tail structure clasped under the body. When threatened by a predator, a springtail can release this and launch itself to safety.

LENGTH: 0.004–0.3 INCH (0.1–8.0 MM)

Lygaeus kalmii (SMALL MILKWEED BUG)

Small milkweed bugs appear dead in the deep cold of winter, but they are usually hibernating, becoming animated enough on warm days to come out and sunbathe.

A QUEEN BEE'S WINTER SOJOURN

QUEEN BUMBLE BEE (Bombus spp.)

This QUEEN BUMBLE BEE is resting in her winter den. She is larger than the worker and drone bees, which do not overwinter.

Once the queen feels the warmth returning to the thatch and soil, she emerges and establishes a new underground nest. Here she begins to store pollen gathered from early-blooming flowers, then begins to lay eggs—for a new generation of bumble bees.

The Living Soil

A few inches deeper, in the thatch layer where dead vegetation is transformed into soil, airy porous spaces provide one of the most abundant and active ecosystems on the planet. This thin soil horizon, this boundary between aboveground and belowground, teems with life, like a miniature terrestrial coral reef.

Digging into this layer, even in midwinter, can reveal whole groups of springtails, different from snow fleas but just as active. Woodlice remain abundant and semiactive in this layer. Careful sifting and sorting can produce a huge range of other tiny animals: some dead, some active, some asleep.

Here, though rarely seen, queen bumble bees dig themselves down into cozy beds of fluffy soil and earthy dead vegetation. They remain groggy and inactive throughout winter, like hibernating bears. Pumped full of their own biochemical antifreeze, they withstand repeated freezing and thawing and the massive pressure of heavy snowpack.

Some bee researchers believe that bumble bee queens select cool, north-facing slopes, the last places to warm in spring. This lets them remain in hibernation until the aboveground world has flowers ready for them.

The thatch layer, a tangle of living and dead plant stems, leaves, and roots, is one of Earth's most interesting and overlooked ecosystems. Within this woven complex exist myriad creatures and organisms—from fungal hyphae to fireflies, garter snakes to singing crickets—as well as strangely parasitic plants such as dodder (*Cuscuta* spp.), which spreads tendrils among live and dead plant tissue, seeking host plants to entangle and draw nutrients from.

Still deeper in the soil, beneath the frost line, grublike larvae of milkweed longhorn beetles (*Tetraopes* spp.) may continue to feed, especially in late winter. Longhorn beetles bore tunnels (sometimes called galleries, as though they might curate art within them) into the center of thick milkweed roots and crowns. There they live and defecate, surrounded entirely by their own food source.

These beetles may also venture beyond their galleries to feed on smaller roots. With less protection, however, they are at risk of attack by moles, or by small insect-feeding nematodes—the latter scenario comparable to a full-grown human being attacked and eaten by a much smaller snake or eel.

Mollisols (from the Latin for "soft soil") are the largest category of North American prairie and grassland soils. Common in the Midwest and Great Plains, they are also widespread in parts of eastern Europe and Asia. Historically, the nutrient-rich surface horizon of North American prairie soils frequently exceeded 3 feet (80 cm) in depth.

Soil, of course, is also interwoven with tangled masses of plant roots, fine and fibrous, thick and woody. The life-filled space inhabited by roots, sometimes called the rhizosphere, is a primary feedstock for countless other living things.

DEEP IN THE PRAIRIE ROOT ZONE

HISTORICALLY, THE UPPER HORIZONS OF NORTH AMERICAN PRAIRIE SOILS OFTEN EXCEEDED 3 FT (80 CM) IN DEPTH

In their larval stage, *Tetraopes tetrophthalmus* (milkweed longhorn beetles) feed belowground on milkweed root systems.

NUTRIENT EXCHANGE
ROOT HAIRS RELEASE AS WELL AS ABSORB NUTRIENTS

minerals

H_2O

sugars

proteins

sugars

proteins

H_2O

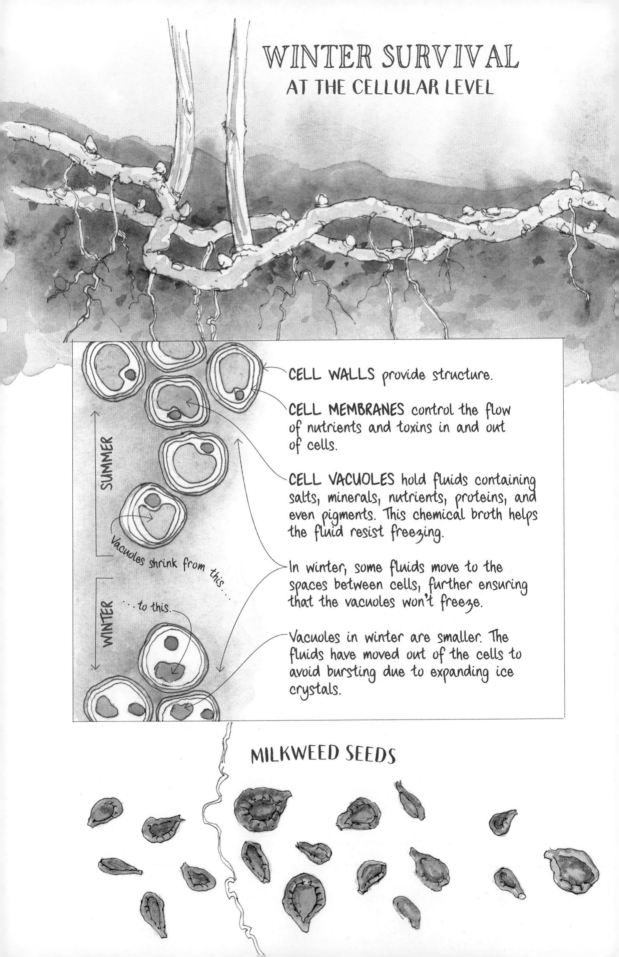

WINTER SURVIVAL
AT THE CELLULAR LEVEL

CELL WALLS provide structure.

CELL MEMBRANES control the flow of nutrients and toxins in and out of cells.

CELL VACUOLES hold fluids containing salts, minerals, nutrients, proteins, and even pigments. This chemical broth helps the fluid resist freezing.

In winter, some fluids move to the spaces between cells, further ensuring that the vacuoles won't freeze.

Vacuoles in winter are smaller. The fluids have moved out of the cells to avoid bursting due to expanding ice crystals.

SUMMER

WINTER

Vacuoles shrink from this...

...to this.

MILKWEED SEEDS

A Plant's Winter Work

Alive and active in these vastly different seasons, plants collect the energy of the summer sun and release it, even in winter, as root secretions of sugars and proteins, as well as allelochemicals that can benefit or deter other organisms living near a plant. In other words, to some significant degree plants choose their neighbors by deciding which microbes they want to attract or repel with biochemical offerings. Those microbes, in turn, attract varying multicellular animals that feed upon them.

While their functions slow in winter, plants (like animals) adapt to the freeze. In common milkweed and other plants, fluids move out of root cells into the surrounding (or intercellular) spaces. Ice crystals may form but less so in the cells themselves, which are protected by fluid with accumulated salts, sugars, and other substances. Like bumble bees and snow fleas, milkweed makes its own antifreeze.

For some milkweeds, the prairie winter itself is the genesis of life. Buried in the subnivean zone (the space beneath the snowpack), seeds from the previous autumn undergo repeated freezing and thawing. These processes degrade the hard outer seed coat and expose the inner embryo to moisture and chemical clues that help prompt spring germination.

Many plants produce seeds that are physiologically dormant—not germinating unless exposed to the long cold of winter. Milkweeds, however, are quite variable.

Mysteriously, different regional populations of the same milkweed species vary in how their seeds respond to temperature. For example, in one location, seeds of showy milkweed plants (*Asclepias speciosa*) germinate most strongly after a prolonged cold spell. Seeds from another location, however, may germinate only slightly better after cold weather. The same species from a third location may need no cold-weather exposure at all. All the plants are still showy milkweed, but the separate populations have evolved to have their own separate traits.

Shelterbelts and Trash Trees

To witness the rich life of winter—the wind-desiccated seedpods, the snow fleas, the food caches of voles—requires a sort of devotion to cold and solitude, a devotion to lonely places, such as windswept shelterbelts. Because of this, the mouse nests and slumbering milkweed bugs go mostly unnoticed.

Many of North Dakota's first shelterbelts were established during the Dust Bowl. Often they were planted with tough species imported from cold Asian deserts—trees that survive in barren and windy places where more delicate species won't. I grew up around remnants of the first shelterbelts planted, derelict rows of such trees as Siberian elm (*Ulmus pumila*). Since childhood, I've heard these described as "trash trees." They frequently die back to the ground from diseases, then produce a profusion of new shoots from still-living underground root systems. Eventually they form thickets tangled with overgrown grass and wild vegetation—including milkweeds.

Seeing these shelterbelts now, with noble tangles of native milkweed contained within them, is to see places full of wild exuberance. Perhaps not the nature we would have imagined, but the nature we have.

WINDSWEPT SPECIES
OF THE NORTHERN GREAT PLAINS

1. *Asclepias syriaca* (common milkweed): native
2. *Ulmus pumila* (Siberian elm): introduced
3. *Shepherdia argentea* (silver buffaloberry): native
4. *Elymus canadensis* (Canada wild rye): native
5. *Daucus carota* (Queen Anne's lace): introduced
6. *Prunus fruticosa* (Mongolian cherry): introduced

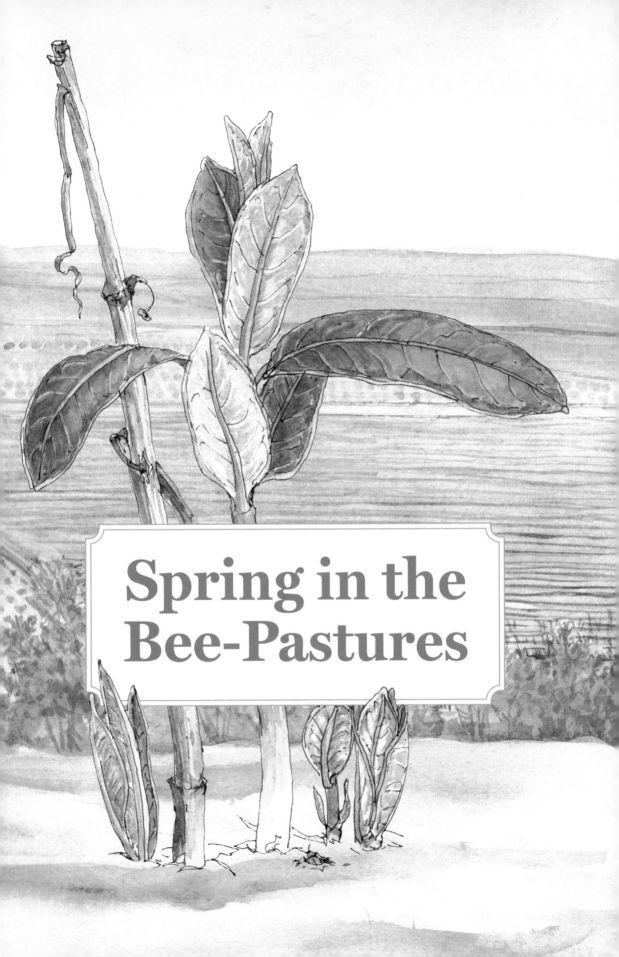

Spring in the
Bee-Pastures

CALIFORNIA'S CENTRAL VALLEY is a land without its native vegetation. Instead, it is a filled and drained former wetland that supports agriculture of a hard-to-grasp scale, along with the occasional weeds that grow around the edges. Milkweeds in particular are conspicuously absent—and seemingly have been for a long time.

This absence is something of a mystery. Herbarium records going back a century show several milkweed species distributed across the state: showy (*Asclepias speciosa*), narrowleaf (*A. fascicularis*), woollypod (*A. eriocarpa*), California (*A. californica*), heartleaf (*A. cordifolia*), and others. All have been found, and amply documented, on mountains and foothills but rarely in the Central Valley itself. Perhaps they had once been there but, being especially sensitive to disturbance, were among the first to vanish when the area was converted to farmland. Or perhaps they never thrived there due to the ancient floods that used to fill the Valley.

A Bed of Honey-Bloom

Nineteenth-century naturalist John Muir described the Central Valley as an incredibly flower-rich landscape. In his 1894 essay "The Bee-Pastures," he wrote:

> *The Great Central Plain of California, during the months of March, April, and May, was one smooth, continuous bed of honey-bloom, so marvelously rich that, in walking from one end of it to the other, a distance of more than 400 miles, your foot would press about a hundred flowers at every step. Mints, gilias, nemophilas, castilleias, and innumerable compositae were so crowded together that, had ninety-nine per cent of them been taken away, the plain would still have seemed to any but Californians extravagantly flowery.*

UNCOMMON IN THE VALLEY

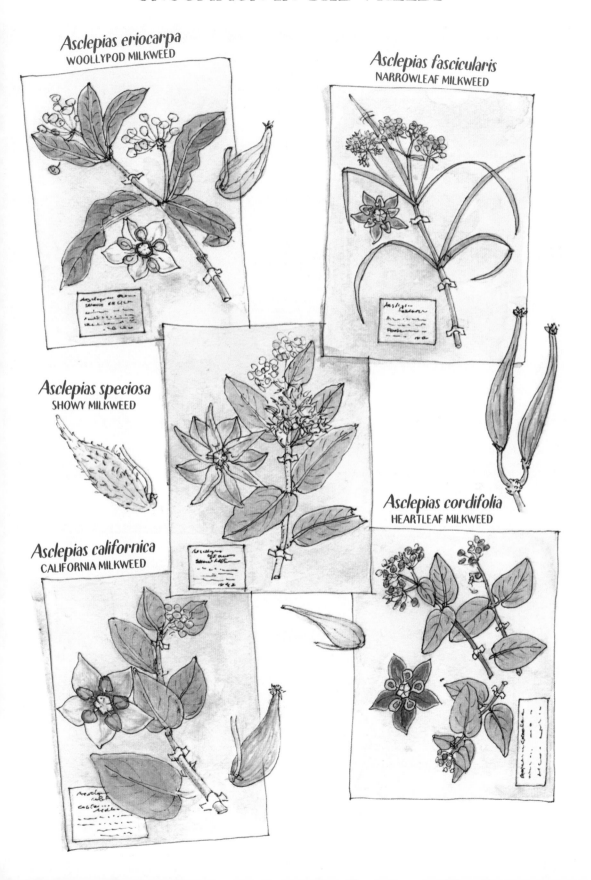

Asclepias eriocarpa
WOOLLYPOD MILKWEED

Asclepias fascicularis
NARROWLEAF MILKWEED

Asclepias speciosa
SHOWY MILKWEED

Asclepias cordifolia
HEARTLEAF MILKWEED

Asclepias californica
CALIFORNIA MILKWEED

A TRANSFORMED LANDSCAPE

Clarkia purpurea
(winecup clarkia)

Gilia achilleifolia
(California gilia)

*Eschscholzia
californica*
(California poppy)

Nineteenth Century: THE BEE-PASTURES OF CALIFORNIA

Oryza sativa
(rice)

Prunus dulcis
(almond)

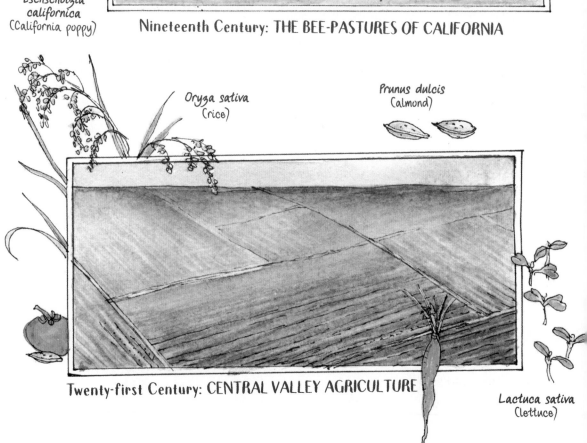

Twenty-first Century: CENTRAL VALLEY AGRICULTURE

Lactuca sativa
(lettuce)

Today almost none of this remains. Possibly the most intensively farmed land on Earth, the Valley is expansively flat with nearly every bump or irregularity plowed into conformity. Enormous machines kick up dust clouds on field roads. Crop dusters rain pesticides from above. Steep, muddy irrigation ditches carry dirty brown liquid across the landscape.

It's a futuristic version of antilife: drought, a teetering climate, and political fights over scarce water; wildfire smoke and high cancer rates; poverty.

It's also where an overwhelming majority of our everyday foods come from: lettuce, almonds, carrots, canned tomatoes, rice, and dozens of other crops. It's a landscape of infinitely decent, kind, and hope-filled families who have managed to make a solidly middle-class existence for themselves in farming, food processing, or trucking—Sikhs, Mormons, Armenians, Mexicans, Japanese, Germans—a colonial and refugee melting pot of our ancestors. The best of us.

Still, it is completely devoid of Old Muir's bee-pastures.

The Alchemy of John Anderson

John Anderson, a veterinarian at the University of California at Davis, had a boundless love of wild animals and nature. In the 1980s, when he and his wife, Marsha, settled onto the dusty, pesticide-laden, scorched-earth farmland of the Central Valley—long California's agricultural powerhouse by then—he was struck by the absence of wildlife.

Over years of steady work, John planted hedgerows of native trees and shrubs along his property edges. Then he began adding patches of grasses and wildflowers.

Incrementally, animals began to return. First were small creatures such as quail and butterflies; then rare songbirds not seen for decades. Eventually the larger beasts came: coyotes, deer, even a bear.

This massive home project became Hedgerow Farms, an acclaimed native plant seed company as well as an informal meeting place and idea hub for people interested in the nuts and bolts of native plant restoration.

Reenter Milkweed

John was a mentor, a friend, and something like a second father to a generation of West Coast conservationists. Mention his name to almost anyone connected with California native plants, and they will tell stories of his picking up rattlesnakes or helping someone return an abused sheep pasture to a state of vegetative glory.

I was part of a small cadre of conspirators who persuaded John to grow milkweed as a seed crop, ensuring a steady supply for conservation agencies to use in habitat restoration projects for monarch butterflies. Through my day job at the Xerces Society we had received grant money to help launch the scheme. Hedgerow Farms would grow the milkweed, and Xerces would help troubleshoot any problems.

To grow the crop, John found plants to collect seeds from near Mount Shasta. The long-distance seed trafficking was necessary because we simply couldn't find any local milkweeds. Those wild plants served as the parents for a newly established seed production field several hours south in the heart of the Central Valley.

REWILDING CALIFORNIA'S
CENTRAL VALLEY
PLANTING MILKWEED AND RESTORING HABITAT

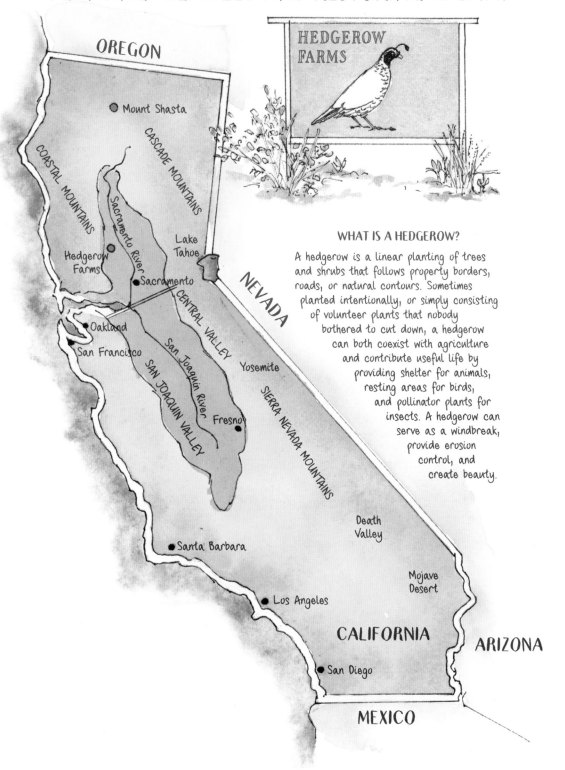

HEDGEROW FARMS

OREGON

Mount Shasta

CASCADE MOUNTAINS

COASTAL MOUNTAINS

Sacramento River

Hedgerow Farms

Lake Tahoe

Sacramento

NEVADA

CENTRAL VALLEY

Oakland

San Francisco

San Joaquin River

SAN JOAQUIN VALLEY

Yosemite

Fresno

SIERRA NEVADA MOUNTAINS

Santa Barbara

Death Valley

Mojave Desert

Los Angeles

CALIFORNIA

ARIZONA

San Diego

MEXICO

WHAT IS A HEDGEROW?

A hedgerow is a linear planting of trees and shrubs that follows property borders, roads, or natural contours. Sometimes planted intentionally, or simply consisting of volunteer plants that nobody bothered to cut down, a hedgerow can both coexist with agriculture and contribute useful life by providing shelter for animals, resting areas for birds, and pollinator plants for insects. A hedgerow can serve as a windbreak, provide erosion control, and create beauty.

The Beetle Plague

Ours was an easy partnership until the beetles appeared, boiling up from the ground like a kind of biblical plague. I pretty much ignored John the first time he called about them. I was too busy with other things. It couldn't possibly be such a big deal. "Lots of things eat milkweed, John. It will be fine."

Two days later, he called again. This time his staff were picking thousands of them off the plants and attempting to drown them in buckets of soapy water. Grudgingly, I booked a flight to go assess the situation.

The cobalt milkweed beetle possesses a stupefying beauty. Glimmering with a deep reflective indigo, it can also appear purple or gold or green depending on the angle and intensity of light reflecting off it. It is hard to look away from, a living, luminous jewel.

The beetles invading Hedgerow Farms had overwintered in the soil, where they had secretly fed upon the milkweed rhizomes. After undergoing metamorphosis from a grublike larval stage, they emerged as adults just as the first spring milkweed shoots broke ground. Then they turned their feeding attention to the aboveground foliage.

THE COBALT MILKWEED BEETLE
Chrysochus cobaltinus

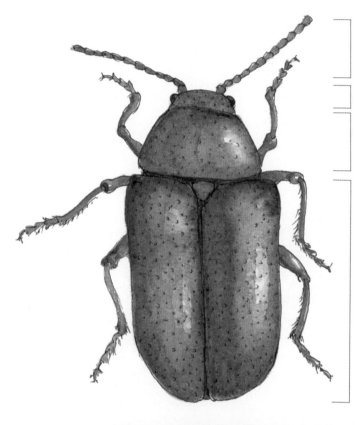

KEY FEATURES

Antennae

Clypeus (the front face "shield")
Eyes

Pronotum (thoracic "shield")

Scutellum (wing base plate)

Spiracles (respiratory openings)

Elytral suture
(opening between wing covers)

Elytra (wing covers)

Femurs

Tibias

Tarsi

Tarsal claws

RANGE OF COLORS

THE COBALT MILKWEED BEETLE:

- Is a medium-size beetle (¼"-¾" or 6-9 mm long)
- Goes through complete metamorphosis, from golden egg to larva to pupa to the bright blue adult
- Feeds (as both larva and adult) aboveground on milkweed plants for up to 6 weeks

- Is one member of the beetle order, Coleoptera, which represents about a third of all animals and 40% of all insects
- Can "launch" itself 2 to 3 times its length to avoid danger

THE COBALT MILKWEED BEETLE
LIFE CYCLE

Most of the visible aboveground activity occurs during 4 to 6 weeks
in late spring and early summer, when adult beetles emerge and mate.
Eggs are then laid on milkweed leaf surfaces.

The hatching larvae drop to the ground and burrow into the soil. There they
feed upon milkweed roots before pupating and emerging the following spring.

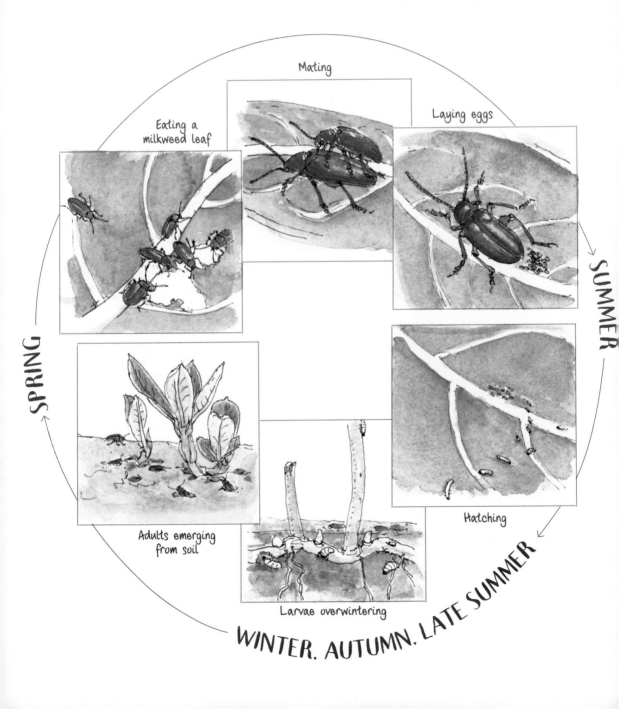

Mating

Laying eggs

Eating a
milkweed leaf

SUMMER

SPRING

Hatching

Adults emerging
from soil

Larvae overwintering

WINTER, AUTUMN, LATE SUMMER

The crop damage at Hedgerow Farms was immense. Affected plants were in only their first year of field production, and although the young milkweed shoots were just a few inches tall, many bore a dozen or more rapidly chewing beetles.

By late spring, the cobalt beetles had run their course. In the end, I think John was in awe of the beetles as much I was. I have vivid memories of him crouched down in the field marveling at them, although it never did stop him from trying to keep their numbers down. And although it was mayhem, the showy milkweeds that grew at Hedgerow Farms were probably the first in a century to grow within a 40-mile (64 km) radius.

The alchemy that John Anderson contributed to this place was that he could conjure animals into existence simply by restoring the land's original plants. The beetles were just one more example. Like the milkweed, they may have been the first of their species in a century to reappear within 40 miles.

Aphids on the Wind

Spring milkweeds don't only summon animals from belowground, but they also invoke life from the winds. I discovered this early in my professional life when the aphids fell to earth one spring day in Wisconsin.

It happened during my first year as a crop consultant for native seed producers. I was a kind of overseer of some hundred species of wildflowers and grasses spread out across two counties and a thousand acres. It was a solitary vocation of long, muddy days in the truck with a thermos and a ream of useless spreadsheets. Ostensibly, I was to monitor the health and condition of the seed crops, although the multitudes of plants would have been just as well off without me.

Early that spring, weeks of southerly winds brought sandhill cranes to forage in the crop fields and bright yellow aphids to infest emerging milkweeds. Oleander aphids (*Aphis nerii*) are an invasive species from the Mediterranean region, where they are closely associated with their namesake plant, oleander (*Nerium oleander*), a showy flowering shrub in the dogbane family.

Born Pregnant

Even among insects, oleander aphids are extraordinary. Rather than laying eggs (as most proper invertebrates do), the females give birth to live offspring. Those offspring are born already pregnant and ready to give birth themselves.

SPRING IN THE BEE-PASTURES

THE OLEANDER APHID

Aphis nerii

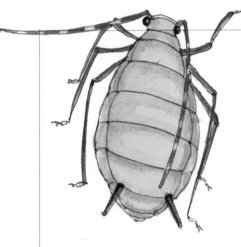

- Native to the Mediterranean region, this aphid is now distributed globally across many temperate, tropical, and subtropical regions.

- Host plants include milkweed, dogbane, oleander, periwinkle, spurge, tobacco, and more.

- In crowded conditions, or when food is scarce, it may develop wings to help it colonize more favorable conditions elsewhere.

PARTHENOGENESIS: ASEXUAL REPRODUCTION

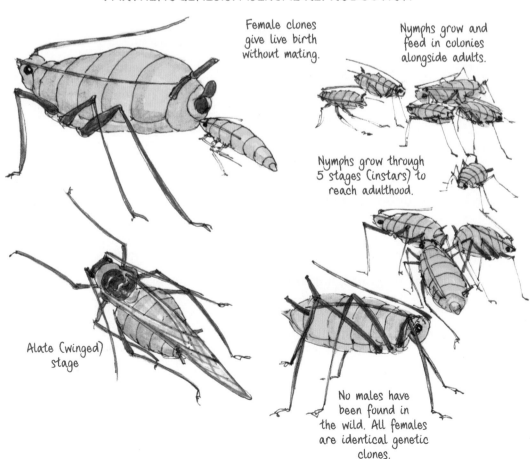

Female clones give live birth without mating.

Nymphs grow and feed in colonies alongside adults.

Nymphs grow through 5 stages (instars) to reach adulthood.

Alate (winged) stage

No males have been found in the wild. All females are identical genetic clones.

More bizarrely, they do this without even mating. Males of this species are not found in the wild. To the best of our knowledge, this makes oleander aphids a species of clones, the Russian dolls of the insect world. Since they can replicate themselves constantly and continuously, their populations can grow explosively when they land on a favorable food source. As it turns out, milkweeds are their top choice.

With this constant birthing, as well as their warm-climate pedigree, oleander aphids don't hibernate, so they need a constant source of food. Thus, they reside as a year-round population along the Gulf Coast, in Florida, Texas, and California, and south into Mexico, Central America, and the tropics. Driven by hunger, opportunity, and crowding, some of these aphids grow wings (a stage known as an alate) and take to the air in search of new food sources.

Inevitably, this brings them northward every spring as winter recedes across the Midwest. There they form new seasonal outposts on every milkweed plant they can find. The closest native equivalent might be *Aphis asclepiadis*, a locally migratory species that lives out its life cycle on multiple plant species, feeding on dogwood leaves in spring before migrating to milkweed and other plants in summer.

Although aphids are poor fliers, their tiny size and light body weight benefit from even modest tailwinds, propelling them into far-flung new places. The wind and sky carry and deposit them constantly across the earth, along with throngs of other diminutive insects, minute seeds, pollen grains, fungal spores, bacteria, and other tiny masses of living things (collectively called aeroplankton). Together with the sandhill cranes, these small wonders are the rightful owners of spring's abundance.

CLONES AND MORE CLONES
FEEDING ON MILKWEED AS A COLONY

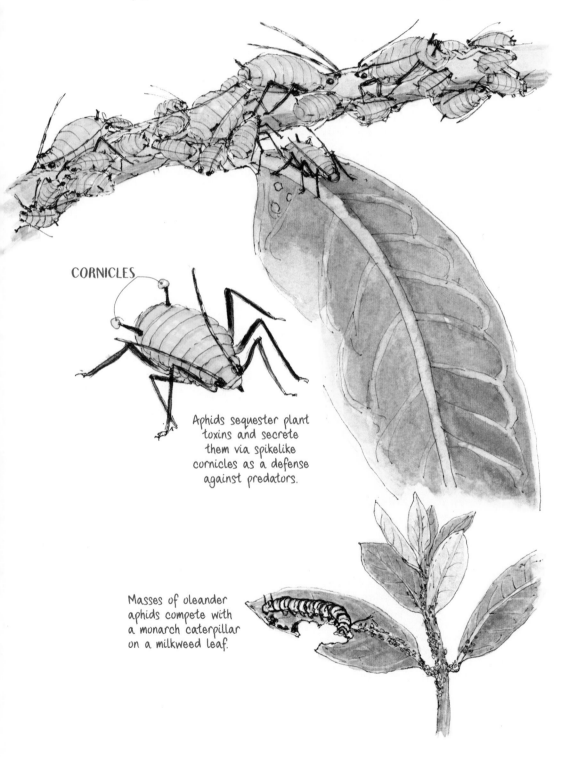

CORNICLES

Aphids sequester plant toxins and secrete them via spikelike cornicles as a defense against predators.

Masses of oleander aphids compete with a monarch caterpillar on a milkweed leaf.

The Vandal Horde

Oleander aphids are a near-constant curse upon milkweed seed producers. In their northward travels the aphids sample redundant countless plants along the way. On these stopovers, they inject their syringelike feeding tubes into plants, regurgitating digestive enzymes, and then they slurp up the resulting goo, like a milkshake of saliva and plant juice. In this process they pick up and spread various plant viruses that cause deformed and discolored foliage, and they defecate undigested sugars that grow foliar molds and rot milkweed leaves.

In a short time, individual stems can become covered by thousands of birthing clones, all literally sucking the life out of a plant. Even monarch butterfly caterpillars are frequently forced to muddle through aphid hordes in search of foliage to eat. I haven't (and to the best of my knowledge, nobody has) quantified how much these nonnative aphids impact milkweed seed crops, but it's hard to imagine that they can't be doing significant harm.

Oleander aphids thrive because, like monarchs or cobalt milkweed beetles, they sequester milkweed cardenolides as a redundant protection against predators. This chemical defense is freely excreted through the cornicles, spike-like tubes projecting from their rumps. The effect is much like growing poison-tipped spears from one's hind end as a defense against sneak attackers. Small spiders that momentarily grasp an oleander aphid are sometimes treated to a quick taste of these secretions, prompting them to abandon their catch and make a hasty retreat.

Thanks to this defense system, oleander aphids, like most other milkweed insects, have evolved to be unfazed by larger animals, choosing to remain in a state of near-constant feeding rather than fleeing

A WISP OF A WASP
A PREDATOR OF THE OLEANDER APHID

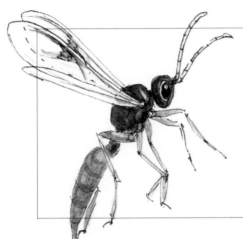

Lysiphlebus testaceipes

At less than ⅛" (3 mm) in length, this member of the Braconidae family feeds on at least 100 species of aphids.

PARASITE OR PARASITOID?

In scientific terminology, parasitoids are a subcategory of parasites that ultimately kill their host organism.

THE BASIC LIFE CYCLE OF PARASITOID WASPS

3. Infected host aphids appear swollen and discolored in the late stages of wasp larval development. Eventually the wasp will secure the newly dead host to a leaf through a silk porthole on the aphid's underside.

5. The hollowed-out corpses of host aphids are marked with prominent wasp emergence holes.

4. After pupation, the adult wasp will chew an exit hole and emerge from the aphid's cadaver.

1. Parasitoid wasps inject eggs into a host using their stinger-like ovipositor.

2. Upon hatching, a wasp larva will feed on the live host from the inside, without killing it, until it is ready to pupate.

potential danger. Look closer, however, and you'll notice that even the prolific and well-protected oleander aphid is not completely without predators.

A Tiny Wasp Enters the Fray

One of these is *Lysiphlebus testaceipes*, a tiny wisp of a creature with no common name. Scarcely visible to the unaided eye at less than one-eighth inch (3 mm) in length, this parasitic braconid wasp seems to be immune to the effects of cardenolides.

Wasps of this type are famed for their otherworldly parasitic life cycle. The female injects a tiny egg into her aphid prey via a stinger-like ovipositor, basically an egg-injecting syringe. The wasp offspring hatches inside the aphid and feeds on its still-living prey, from within, over the course of several days. Eventually it kills the aphid in the final act of feeding upon it, like the closing scene of some macabre play.

Next the wasp larva bites a hole in the bottom of the aphid and glues its carcass to the leaf surface with silk, lest a strong wind blow away the wasp's cadaver home. Within this mummified tomb, the wasp pupates for several days. Then, upon reaching adulthood, it chews a perfectly round exit hole in the aphid's back and departs.

Aphids in the stages of this parasitism turn brown and appear slightly bloated, and if you look attentively at a big colony, you can see these mummies. Those with exit holes are even more conspicuous. Their corpses seem to have been pierced with a perfect miniature paper punch.

Patient, close-up observations of big aphid colonies will often reveal these nearly microscopic wasps as they flit daintily among prey.

MYRMECOPHILY
A PARTNERSHIP OF ANTS AND APHIDS

The term myrmecophily ("love of ants") describes mutualistic (and sometimes parasitic) partnerships between ants and other species.

Sticky, sugary aphid excrements will attract ants.

APHID TENDING BY ANTS

An ant sips a drop of aphid waste.

The aphid's sugary excrement provides a high-value food source to ants.

Like tiny shepherds, ants watch over aphid colonies, warding off predators.

Herded by Ants

Some aphid species are tended by ants, like livestock being watched over by shepherds. One of these is *Aphis asclepiadis*, a common aphid of milkweeds. In these widely observed relationships—known as myrmecophily—ants stand guard over their aphid flock, fighting off predators and even removing diseased aphids from a plant before they can infect others.

Because aphids primarily digest plant proteins and defecate plant sugars (or "honeydew"), they provide a constant source of sugary food for the ants, who dutifully collect (or "milk") it and carry it back to their nest.

From Oasis to Monoculture

Although shockingly prolific in a milkweed seed field, aphids, beetles, and parasitic wasps are less common in the Upper Midwest, where large sections have been transformed into a simplified landscape of corn and soy and not much else.

As recently as three or four decades ago, it was much more varied and diverse, with dairies that had ample weedy pastures, vegetable farms growing peas and green beans, the occasional truck farm growing strawberries or pumpkins, and woodlots for hunting and firewood. Beyond this, 30 million acres of land were enrolled into the USDA's Conservation Reserve Program, which paid farmers to maintain highly erodible lands as permanent grassland rather than try to grow crops on them. Those land uses allowed milkweeds and other native plants to persist in the fringes.

The corn and soy empire has largely replaced those old farming systems. It also relentlessly expands westward, northward, ever farther beyond the boundaries that formerly contained it. It's an empire bene-fiting from a warming climate and ever-faster-maturing varieties.

Both crops are genetically modified, altered for herbicide resis-tance. They flourish after spraying even while the weeds around them die. They are planted as Day-Glo seed pellets, coated in insecticide dust and dyed fluorescent colors to warn of their toxicity.

A Wild Idea

Like Hedgerow Farms, the Midwest native seed producers invented an oasis of wildness in the middle of status quo agriculture. Along with oleander aphids, their fields drone and bustle with odd prairie birds like the dickcissel. The wet spring ground of these milkweed lands comes alive with firefly larvae hunting for slugs in the leaf litter. Skunks appear with kits. And more than once I have stumbled upon snapping turtles that clambered hundreds of feet out of nearby ponds to lay their eggs among the swamp milkweed, sneezeweed, and prairie blazing star.

A little bit of that wildness gets packaged up and sent off to cus-tomers, who buy the seed and plant it—who knows where—around suburban houses, city parking strips, or wastewater treatment ponds.

A RECONSTRUCTED NATURE
A CENTRAL VALLEY HEDGEROW

1. *Asclepias speciosa* (showy milkweed)
2. *Sambucus nigra subsp. caerulea* (blue elder)
3. *Eschscholzia californica* (California poppy)
4. *Clarkia spp.* (clarkia)
5. *Muhlenbergia rigens* (deergrass)
6. *Ceanothus spp.* (California lilac)
7. *Regulus calendula* (ruby-crowned kinglet)
8. *Callipepla californica* (California quail)
9. Almond orchards . . . approximately 900,000 acres in California's Central Valley

The Hedgerow Way

Back in California, before he was a seed producer, John Anderson was one of the first people to take the arcane and mostly European practice of hedgerow planting and to reinterpret it for his own surroundings. He did this with long arrays of native trees and shrubs along his property lines: elderberry, ceanothus, coffeeberry, and a dozen others. The hedges were framed by understory plantings of deergrass, lupines, poppies, clarkia, and a richly textured mix of other native grasses and wildflowers. Hence the name of his farm.

Since then, other stakeholders have built onto this template. Hedgerows have become one of the few great conservation success stories in the Central Valley. The Xerces Society, Wild Farm Alliance, California Audubon, and numerous government conservation agencies have threaded hundreds of miles of hedgerows across an otherwise parched, dusty, industrial land. Most often these are the only natural habitat for many miles, and many contain the offspring of John's original showy milkweed plants, the crop that survived the year of the beetles.

I still travel to the Central Valley to check on these hedges for work, navigating desolate, unnamed country roads between almond orchards that all look alike. My favorite hedges, sometimes a mile or more in length, are the ones that don't look the best. They have weird gaps in them, or beget unexpected weeds such as stinging nettle, or gather bits of trash from the nearby roads.

Even within these rough hedges there's always some odd collection of butterflies and quail, coyotes and snakes, beetles and aphids. They are animals that by some miracle wandered in from the surrounding toxic moonscape and made a home.

Summer
Currents

SUMMER MILKWEEDS ARE A RIOT of life, prosperity, and struggle. Few other plant groups suffer the vast legions of frenetic herbivores that plague summer milkweeds. At the same time, summer is when milkweeds must rally as much energy as possible to make bold flowers and attract pollinators—the go-betweens for milkweed sex.

Roadside Ecology

Among my people, common milkweed is a "ditch weed." It grows with other ditch weeds such as smooth bromegrass, sheep sorrel, and Queen Anne's lace. Somehow it persists along roadsides where the prairie aristocrats, blazing stars and pasqueflowers, died out a century ago.

There's a gritty, perverse perseverance to highway ditches and ditch ecology. On one side there's often some ecologically impoverished crop field. Often these are sprayed with miracle chemicals: dicamba, glyphosate, 2,4-D amine, atrazine, metolachlor, sulfentrazone, flumioxazin, pendimethalin, azoxystrobin, bifenthrin, and clothianidin. On the other side is the constant whoosh of interstate highway commerce. Trucks produce enough air turbulence to nearly knock you over, while local cars speed on toward distant destinations.

Although narrow and sad, ditches can extend unbroken for miles, providing the only seminatural vegetation in some places. Common milkweed arises in ditches like an object out of time, an ancient chemical beacon for wild things.

DITCH WEEDS

Deschampsia cespitosa

Asclepias syriaca

Bromus inermis

Solidago canadensis

Daucus carota

Cirsium discolor

Verbascum spp.

Verbesina alternifolia

Prunella vulgaris

Rumex crispus

Rumex acetosella

MILKWEED FEEDERS

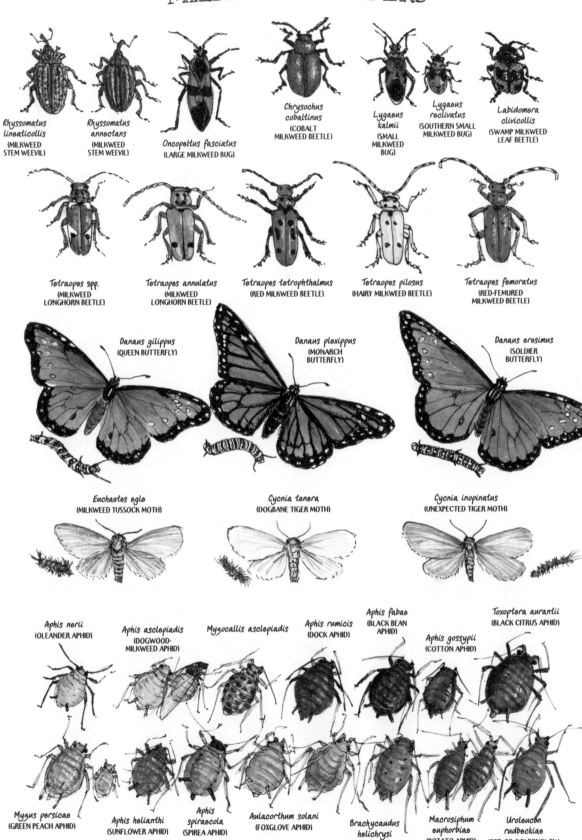

Rhyssomatus lineaticollis (MILKWEED STEM WEEVIL)

Rhyssomatus annectans (MILKWEED STEM WEEVIL)

Oncopeltus fasciatus (LARGE MILKWEED BUG)

Chrysochus cobaltinus (COBALT MILKWEED BEETLE)

Lygaeus kalmii (SMALL MILKWEED BUG)

Lygaeus reclivatus (SOUTHERN SMALL MILKWEED BUG)

Labidomera clivicollis (SWAMP MILKWEED LEAF BEETLE)

Tetraopes spp. (MILKWEED LONGHORN BEETLE)

Tetraopes annulatus (MILKWEED LONGHORN BEETLE)

Tetraopes tetrophthalmus (RED MILKWEED BEETLE)

Tetraopes pilosus (HAIRY MILKWEED BEETLE)

Tetraopes femoratus (RED-FEMURED MILKWEED BEETLE)

Danaus gilippus (QUEEN BUTTERFLY)

Danaus plexippus (MONARCH BUTTERFLY)

Danaus eresimus (SOLDIER BUTTERFLY)

Euchaetes egle (MILKWEED TUSSOCK MOTH)

Cycnia tenera (DOGBANE TIGER MOTH)

Cycnia inopinatus (UNEXPECTED TIGER MOTH)

Aphis nerii (OLEANDER APHID)

Aphis asclepiadis (DOGWOOD-MILKWEED APHID)

Myzocallis asclepiadis

Aphis rumicis (DOCK APHID)

Aphis fabae (BLACK BEAN APHID)

Aphis gossypii (COTTON APHID)

Toxoptera aurantii (BLACK CITRUS APHID)

Myzus persicae (GREEN PEACH APHID)

Aphis helianthi (SUNFLOWER APHID)

Aphis spiraecola (SPIREA APHID)

Aulacorthum solani (FOXGLOVE APHID)

Brachycaudus helichrysi (LEAF-CURLING PLUM APHID)

Macrosiphum euphorbiae (POTATO APHID)

Uroleucon rudbeckiae (RED OR GOLDENGLOW APHID)

The Hungry Throng

By my count at least 40 insects feed often or exclusively on North American milkweeds in the summer: monarch, queen, and soldier butterflies (*Danaus plexippus*, *D. gilippus*, and *D. eresimus*), milkweed tussock moths (*Euchaetes egle*), tiger moths (*Cycnia tenera* and *C. inopinatus*), large milkweed bugs (*Oncopeltus fasciatus*), small milkweed bugs (*Lygaeus kalmii* and *L. reclivatus*), 14 odd species of longhorn beetles (*Tetraopes* spp.), cobalt milkweed beetles (*Chrysochus cobaltinus*), the swamp milkweed leaf beetle (*Labidomera clivicollis*), milkweed stem weevils (*Rhyssomatus lineaticollis* and *R. annectans*), oleander aphids (*Aphis nerii*), dogwood-milkweed aphids (*Aphis asclepiadis*), dock aphids (*Aphis rumicis*), black bean aphids (*Aphis fabae*), cotton aphids (*Aphis gossypii*), spirea aphids (*Aphis spiraecola*), sunflower aphids (*Aphis helianthi*), black citrus aphids (*Toxoptera aurantii*), foxglove aphids (*Aulacorthum solani*), the green peach aphid (*Myzus persicae*), the leaf-curling plum aphid (*Brachycaudus helichrysi*), potato aphids (*Macrosiphum euphorbiae*), red or goldenglow aphids (*Uroleucon rudbeckiae*), and various aphids with no common names (such as *Myzocallis asclepiadis*). Undoubtedly, there are other milkweed insects, undiscovered, or that I am forgetting.

That list of insect species is long, but it fails to express the degree to which these creatures show up in force. A single plant may be burdened with hundreds of diverse insects, each siphoning nutrients out of puncture holes or devouring many leaves; some chew off enough roots to cause entire plants to flop limply onto the ground. Then there are slugs and snails (which, of course, are mollusks, not insects) and spider mites (which are arachnids).

This tally excludes the multitudes of flower-visiting butterflies and moths: fritillaries, swallowtails, skippers, admirals, and blues. There are thousands of bee species ranging from giant carpenter bees nearly as large as a human thumb to sweat bees less than ¼ inch (6 mm) in length. Milkweed nectar also attracts giant, scary-looking solitary wasps with incredible stings, such as the cicada killer and the tarantula hawk. These wasps could floor a person with their venom but are slow to be provoked and graceful, with a quick nimbleness reminiscent of ballet dancers.

Predators of the myriad milkweed insects also abound, such as the (apparently cardenolide-immune) black-headed grosbeak (*Pheucticus melanocephalus*), which sometimes devours monarchs in large numbers, and ghostly white crab spiders that hide among milkweed flowers. Crab spiders align themselves perfectly with flower structures to be camouflaged in plain sight. They are a mousetrap for bees.

A reckoning of summer milkweed feeders must include the various rabbits, jackrabbits, and ground squirrels that, also apparently not susceptible to cardenolide poisoning, gnaw at stems. It's possible, though, that they experience a kind of hallucinatory trip.

All of this life can exist in a ditch. Or not, because ditches can also get mowed.

Wisdom has it that we should mow ditches every year to "clean up" or "keep the bugs down" or "stop weeds from going to seed." Mowing may happen early, before milkweed can even flower. These become calamity years when all the ditch animals flee or die.

Mowing would be easy to postpone if it were not so central to Midwestern identity. Even now, while the prosperity of rural places

MILKWEEDAPOLIS
RESIDENTS AND VISITORS

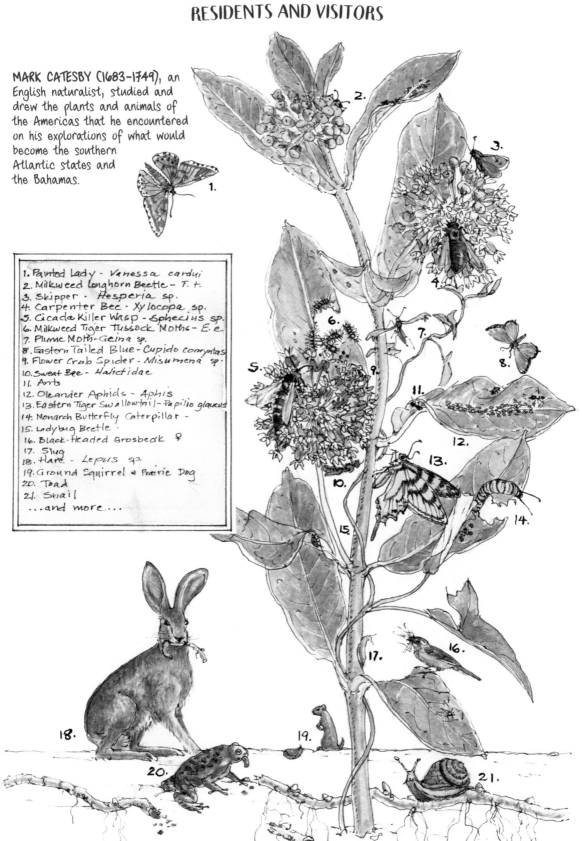

MARK CATESBY (1683-1749), an English naturalist, studied and drew the plants and animals of the Americas that he encountered on his explorations of what would become the southern Atlantic states and the Bahamas.

1. Painted Lady - Vanessa cardui
2. Milkweed Longhorn Beetle - T. t.
3. Skipper - Hesperia sp.
4. Carpenter Bee - Xylocopa sp.
5. Cicada Killer Wasp - Sphecius sp.
6. Milkweed Tiger Tussock Moths - E. e.
7. Plume Moth - Geina sp.
8. Eastern Tailed Blue - Cupido comyntas
9. Flower Crab Spider - Misumena sp.
10. Sweat Bee - Halictidae
11. Ants
12. Oleander Aphids - Aphis
13. Eastern Tiger Swallowtail - Papilio glaucus
14. Monarch Butterfly Caterpillar -
15. Ladybug Beetle -
16. Black-Headed Grosbeak ♀
17. Slug
18. Hare - Lepus sp.
19. Ground Squirrel & Prairie Dog
20. Toad
21. Snail
...and more....

SPECIALIZATION AND SURVIVAL
NICHE HABITATS AND LOCATIONS

Asclepias exaltata
POKE MILKWEED
2-5 feet (60-175 cm) tall

Asclepias subulata
RUSH MILKWEED
3-5 feet (105-175 cm) tall

Asclepias pedicellata
SAVANNAH
MILKWEED
6-12 inches
(15-30 cm) tall

collapses, we still make sure the ditches get mowed. Living a thousand miles away from my homeland, I grapple with this question every year: Should I mow my ditch? Not to do so seems disloyal.

Milkweeds that stay out of ditches have it easier. As well as avoiding the mower, milkweeds that hide themselves can avoid some of their piercing, sucking, and chewing insect companions.

Searching Out the Secretive Milkweeds

In contrast to common milkweed, many other species have found ways to obscure themselves in woods or vast wetlands. Some just stay low to the ground, concealed by grasses, like the savannah milkweed (*Asclepias pedicellata*), a delicate, yellow-flowered plant of Florida that never gets much taller than a foot in height. Rush milkweed (*Asclepias subulata*) might be visited by fewer herbivores simply by growing in the extreme Mojave Desert heat, alongside Joshua trees. Its nearly leafless form resembles desert succulents like cacti.

Often you can find these reclusive milkweeds just by knowing their haunts and associates. In the rich soils of shady Midwest and Appalachian woodlands, you may find poke milkweed (*Asclepias exaltata*), a tall and gangly plant. It keeps company with other massing forbs like white snakeroot (*Ageratina altissima*) under canopies of maple and sycamore. In these forest assemblages, poke milkweed appears less harassed by insects but never really thrives. Where it does venture out of the forest edges, it can hybridize with common milkweed, producing something of an intermediate form.

Swamp Milkweed: The Marvel of Muck

Of these hide-and-seek milkweeds, the one I know best is swamp milk-weed (*Asclepias incarnata*). As the common name would suggest, it's a muck plant, a milkweed I discovered as a kid when my newly single mom packed us up in a Datsun hatchback and drove out of North Dakota to a basement-apartment life in a small Minnesota town on the Mississippi River. Once there, I lived out much of my childhood in mucky places.

Between Minneapolis to the north and St. Louis to the south, the Upper Mississippi River isn't a single river but rather a braided collec-tion of a thousand streams. A deeper main channel constantly breaks up into smaller side channels. Some of these dead-end in shallow, muddy sloughs. Others continue on, reconnecting miles later with the main channel.

Within this are thousands of uninhabited islands, some barely above the water. Thousands of acres of this landscape can disappear in spring floods and enormous ice floes. As a boy, I observed that you could gauge the height of prior flood years by looking up to see how high old plastic bags, dried grass, tires, and other detritus were sus-pended in the cottonwood trees.

True to its name, swamp milkweed prospers in all of this. Possibly more than other members of the milkweed nation, this species is an insect magnet. When gardeners transplant this wetland species onto high ground, I find it covered in longhorn beetles. In these dry gardens, scores of beetle larvae belowground will even remove enough roots to topple the plant.

ALONG THE MISSISSIPPI

Below the headwaters in northern Minnesota, the Upper Mississippi River (between Minneapolis and the confluence with the Ohio River) is a braided, multi-channel water system. Here the river contains countless shifting islands and backwaters.

Despite an abundance of natural habitat, the river is greatly altered by dams, dredging, barge traffic, agricultural runoff, municipal wastewater, and the accumulation of commonplace litter and debris.

Yet in swamp milkweed's natural habitat, just inches above the waterline, longhorn beetles never seem abundant on the plant. To survive in such a place, their larvae would need somehow to withstand not only saturated soils but also icy floods and shifting land. These same challenges may keep other herbivores in check as well, such as milkweed bugs, which normally overwinter in dry leaf litter.

A 2018 research article by scientists at Iowa State University backs up my own observations. The team compared the number of eggs laid by monarchs on eight different milkweed species. When the butterflies were offered a range of milkweed species to choose among, they laid the most eggs on swamp milkweed. When offered only one milkweed species at a time, the butterflies still laid the most eggs on swamp milkweed.

Despite this, monarch caterpillars face the unique risk of floodwaters inundating the wet places where swamp milkweed grows. This is especially hazardous when they are ready to pupate, a period in their development when the caterpillars sometimes climb down from their natal milkweed, seeking a more discreet location to form a chrysalis.

Migration Highway

Fortunately for monarchs, the river floods mostly in spring—before the monarch breeding season, at least in the North. Because of this, the swamp milkweeds of the Upper Mississippi River could be especially important to the monarch migration. As monarch populations have dwindled, we tend to focus on the milkweeds we see from our cars, the ditch weeds. But, largely hidden from view, the Mississippi is its own kind of highway for animal migration.

SWAMP MILKWEED

At heights of close to 6 feet (2 m), and with multiple stems, the surface area of *Asclepias incarnata* provides a great deal of cover and concealment for monarch eggs and caterpillars.

THE CENTRAL FLYWAY
FROM THE GULF OF MEXICO TO NORTHERN MINNESOTA

CANADA

CATFISH

Missouri River

Mississippi River

USA

WOOD DUCK

Ohio River

BLUE TEAL

Mississippi River

BUFFLEHEAD

Atlantic Ocean

MEXICO

Pacific Ocean

Gulf of Mexico

BULL SHARK

The Missouri, Mississippi, and Ohio rivers directly link ecosystems as diverse as high plains, prairies, boreal forests, and the Appalachian mountains with subtropical wetlands and coastal grasslands.

The high ground along the upper river is an unglaciated landscape of steep hills, deep valleys, and sandstone cliff outcroppings, much of it too rough or too steep for farming. The floodplain, which can be miles wide, has flat, rich, silty soils, but floodplain farming is a gamble. Collectively, the wooded bluffs, low forests, grassy islands, cutoff backwater lakes, and interlocking side channels form a large semi-wild north–south corridor in the middle of North America.

Overlay this 800-mile-long (1,300 km) corridor with the migratory path of monarch butterflies heading north every summer. In terms of contiguous breeding habitat and less direct exposure to pesticides, the river valley probably offers better survival odds than anywhere a hundred miles (160 km) to either the east or the west. It doesn't hurt that this corridor also hosts their favorite milkweed and that at least some competing milkweed herbivores get washed away before the monarchs' arrival.

For all this, the river is mostly ignored except among bird-watchers and bird hunters. By one estimate, 40 percent of all North American waterfowl and shorebirds use this migratory route. The same backwaters where I first discovered swamp milkweed were known for the dozens of waterfowl hunters who died in the 1940 Armistice Day blizzard, when the constant flow of ducks overhead tempted locals to stay out until it was too late to find shelter.

To support the famed waterfowl migrations, more than 240,000 acres (27,000 ha) of land along the upper river has been designated as a national wildlife refuge. Aside from those in Florida, this makes it the largest US wildlife refuge east of the Rocky Mountains.

SWAMP MILKWEED'S RIVER COMPANIONS
DIVERSE FISH OF THE MISSISSIPPI SYSTEM

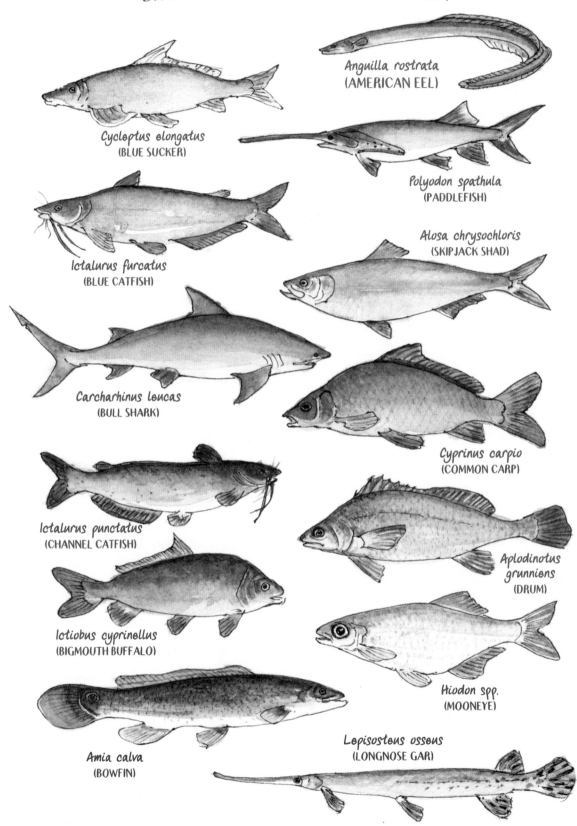

Cycleptus elongatus
(BLUE SUCKER)

Anguilla rostrata
(AMERICAN EEL)

Polyodon spathula
(PADDLEFISH)

Ictalurus furcatus
(BLUE CATFISH)

Alosa chrysochloris
(SKIPJACK SHAD)

Carcharhinus leucas
(BULL SHARK)

Cyprinus carpio
(COMMON CARP)

Ictalurus punctatus
(CHANNEL CATFISH)

Aplodinotus
grunniens
(DRUM)

Ictiobus cyprinellus
(BIGMOUTH BUFFALO)

Hiodon spp.
(MOONEYE)

Amia calva
(BOWFIN)

Lepisosteus osseus
(LONGNOSE GAR)

Rough Fish

Along with birds, the maze of river channels probably once supported migrations of fish as large and dramatic as any Western salmon river. The blue sucker (*Cycleptus elongatus*), paddlefish (*Polyodon spathula*), blue catfish (*Ictalurus furcatus*), and skipjack shad (*Alosa chrysochloris*) most likely made immense runs from Louisiana to Minnesota before the lock and dam systems were built on the upper river.

American eels (*Anguilla rostrata*) certainly came even farther, originating in the Sargasso Sea, then migrating into thousands of tributaries, even slithering out across wet ground on rainy nights in search of distant ponds that they might inhabit for decades before making the return trip. Before the dams, bull sharks were known to migrate from the Gulf of Mexico upriver as far as Illinois.

Although the dams have curtailed the great migrations, river fish— and a boyhood obsession with catching and examining them—are how I met the swamp milkweed. Pushing through a bankside thicket of it on a summer day, I would make my morning commute to a muddy river slough that offered up carp, channel catfish, drum, mooneyes, bullheads, bigmouth buffalo, and a most fascinating air-breathing, biting, cobralike fish that lived in mere inches of brown sludge water, the bowfin.

Referred to as rough fish, they are the equivalent of ditch weeds, survivors of a river laden with farm runoff, pesticides, PCBs, dioxin, mercury, endocrine disruptors, and urinated-out pharmaceuticals. Discovering these beautiful, otherworldly fish in the toxic waters, amid the regal glory of summer swamp milkweed, was a kind of Technicolor childhood experience, a secret church of mud, butterflies, and pink flowers.

A Most Intricate Flower

Flowers are the pinnacle of summer for milkweeds, and, true to character, milkweed makes flowers that diverge into their own extraordinary brand of weirdness.

Among the plant kingdom, flower structures range from simple to complex. Most of us recall this from grade school biology, the simplified illustrations of tuliplike flowers with their stamens and styles, the male and female sex organs, respectively. Depending on the species, these may be combined with an elaborate perianth of colorful petals or large sepals. A flower may or may not produce nectar. A plant may have separate male and female flowers or ones that contain both sex organs. Flowers may grow singly or in complex compound arrangements. All of this is fairly fundamental botany and easy enough to grasp.

Milkweed flowers exist on a different plane. Despite their pedigree as ditch weeds, muck plants, or odd botanical outcasts of harsh and dismal places, even the most commonplace milkweed flowers have a complexity comparable to that of rare orchids. They are among the most elaborate flowers in the plant kingdom.

Pollinator Entanglements

Among the notable features of milkweed flowers are the pollinia (singular: pollinium). These sacs of waxy pollen arise in a ring around the base of the gynostegium. Few other plant groups deliver their pollen this way—orchids being an exception. Most other flowering plants produce individually dispersed grains of pollen.

MILKWEED FLOWER MORPHOLOGY

A common milkweed flower cluster (or umbel)
consists of multiple individual flowers that
arise on short stems from a single branch.

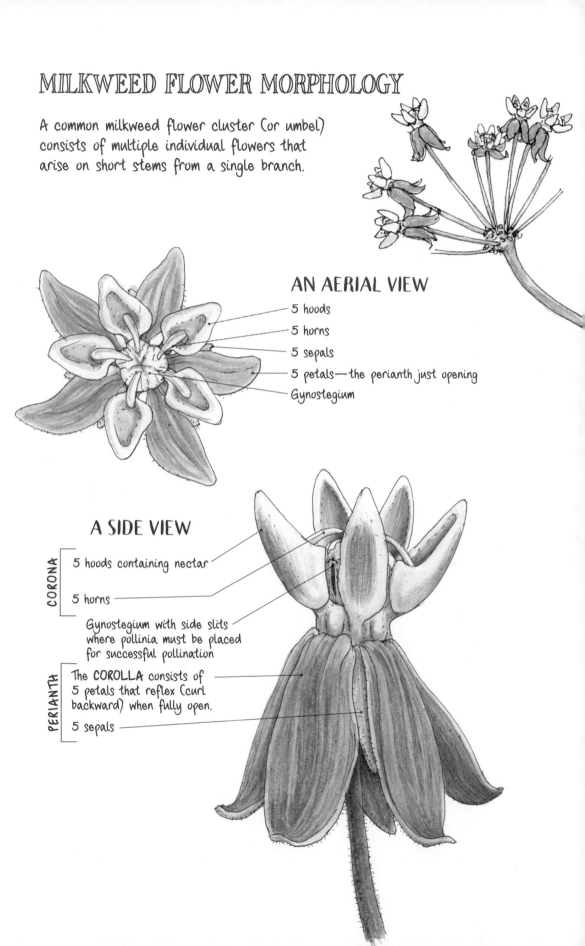

AN AERIAL VIEW

5 hoods

5 horns

5 sepals

5 petals—the perianth just opening

Gynostegium

A SIDE VIEW

CORONA

5 hoods containing nectar

5 horns

Gynostegium with side slits
where pollinia must be placed
for successful pollination

PERIANTH

The COROLLA consists of
5 petals that reflex (curl
backward) when fully open.

5 sepals

A CLOSER LOOK
AT MILKWEED POLLINATION

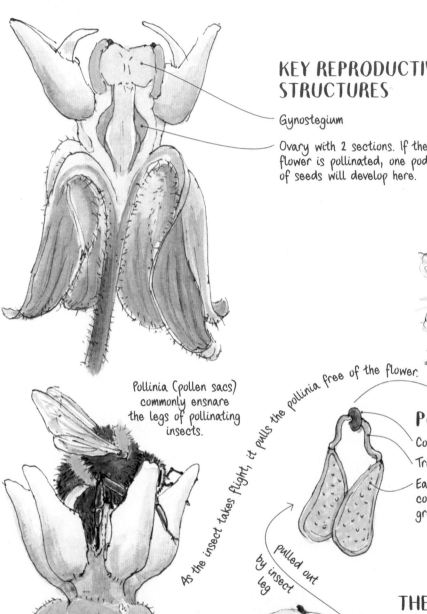

KEY REPRODUCTIVE STRUCTURES

Gynostegium

Ovary with 2 sections. If the flower is pollinated, one pod of seeds will develop here.

Pollinia (pollen sacs) commonly ensnare the legs of pollinating insects.

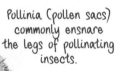

As the insect takes flight, it pulls the pollinia free of the flower.

POLLINIA

Corpusculum

Translator arm

Each sac (pollinium) contains about 100 pollen grains

pulled out
by insect
leg

THE GYNOSTEGIUM

Pollinia showing corpusculum that can catch an insect leg

Pollen sacs

Side slit where pollinia are inserted for pollination

Milkweed pollinia occur in pairs, joined by a sort of wishbone configuration of filaments called translators and a structure called a corpusculum. This acts as a trip wire or trap, attaching itself to the legs or tongues of insects that probe milkweed flowers for nectar.

Bees and wasps, which tend to have numerous small, complex leg spines and claws, are the most frequent bearers of this burden. Small bees can be trapped by the corpusculum and die while stuck to a flower, unable to pull themselves free. Other bees or wasps that make repeated trips to milkweed flowers can have numerous pollinia attached to their feet, making walking cumbersome. Bumble bees can end up with pollinia affixed like binder clips to their tongues.

As an insect travels among flowers, the translator arms of pollinia dry and change shape, causing the sacs to flex outward. This slight shape change creates a sort of key that conveniently and perfectly fits into a milkweed flower's stigmatic slits, located concentrically around the sides of the gynostegium, directly between the slots that produce pollinia.

The insect lands on a new milkweed flower and rotates the pollinia 90 degrees. The key slips into a lock, the insect is relieved of its burden—and a milkweed flower is pollinated.

A fruit, or follicle (the pod packed with seeds), is the end result of a single delicate removal and replacement of pollinia between two milkweed plants.

The odds of all of this happening just so are seemingly astronomical, although the configuration of the flower, including the horn structures, helps guide the footwork of insects so that pollination will

FROM BUD TO POD

A TIMELINE

JUNE 28. A cluster of flowers is visited by potential pollinators.

JULY 12. Unpollinated flowers have dried up and dropped. The pollinated flowers form tiny, swollen, podlike nodules (immature fruits).

JULY 18. More flowers are dropping. Two stems have thickened, with bigger pods.

JULY 20. Two successfully pollinated fruits continue to develop, reaching more than ½ inch (1.4 cm) in length.

AUGUST 13. The two pods are maturing, with up to 100 seeds inside each.

succeed. Moreover, since pollinia take some time to change their shape before fitting into a stigmatic slit, an insect has time to fly around and find an unrelated plant to cross-pollinate.

Still, we know that milkweed pollination is tricky enough business that many flowers seemingly go unpollinated and never make seed. Examining the marvelous pom-pom flower heads of common milkweed, which can contain many dozens of individual flowers, we discover that it's common for only a single one in the cluster to actually be pollinated.

On the other hand, because a single pollinium sac may contain a hundred pollen grains, and a single milkweed flower's ovary may contain a similar number of ovules, the difficult business of milkweed pollination needs to happen only a few times for a plant to create a tremendous number of seeds.

Milkweeds seem to be variable in how much they need cross-pollination between unrelated individuals. Some are thought to have a high degree of self-incompatibility, refusing to be pollinated by their own pollinia. Others seem a little more tolerant of this. In either case, where milkweeds are lost to mowing, herbicides, and habitat fragmentation, the pollination linkages may break down.

Monarchs as Flower Visitors

Butterfly experts note that although adult monarchs are frequent visitors to milkweed flowers, drinking the rich, abundant, and fragrant nectar, they are considered poor milkweed pollinators. Their long, smooth, delicate legs and tongues are hard for the corpusculum to grab. They are probably rarely troubled with pollinia attached to their bodies.

In contrast, a bee pollinator, especially a strong, large bumble bee, is a steadier and more reliable partner for milkweeds in this enterprise. A bumble bee may pick up many pollinia and visit all the other milkweeds in a local area, moving a lot of genetic diversity among the plants within its foraging range. Still, the bumble bee returns home to her nest at night and never really ventures beyond her foraging radius of a mile or so.

In this sense, perhaps we are overlooking something unique about monarchs as pollinators, which is that they fly great distances. With pollinia clamped to the middle tarsus of its leg, a monarch may fly dozens of miles in a day, resulting in the potential long-distance movement of pollinia. This is probably an infrequent occurrence, but possibly significant. It could be that monarchs are functioning as a kind of airborne river for genetic exchange among distantly located plants.

We always knew that monarchs need milkweed. Wouldn't it be interesting if, in turn, milkweeds need monarchs?

SHORT TRIPS AND LONG HAULS

DIFFERENT INSECTS PROVIDE DIFFERENT POLLINATION SERVICES

Short-distance fliers distribute pollinia within the neighborhood.

Long-distance fliers may carry few pollinia . . .

. . . but they carry them afar.

ESTABLISHED MONARCH ROUTES
ALONG THE MILKWEED TRAILS

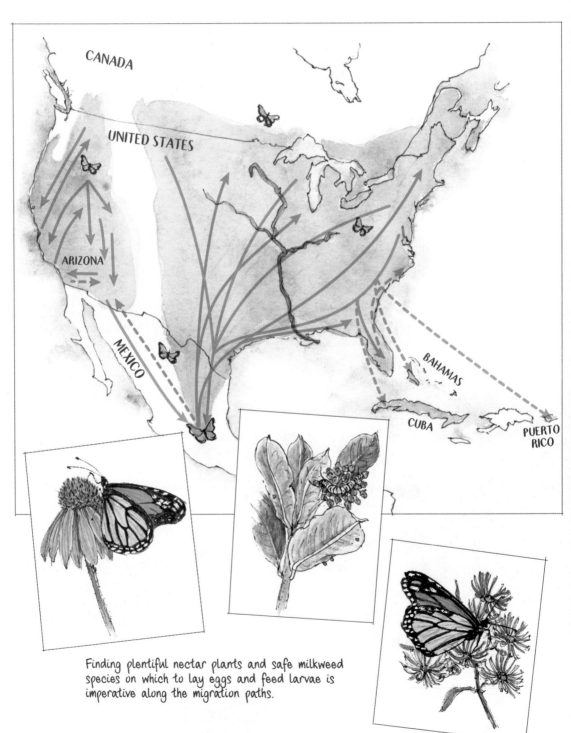

Finding plentiful nectar plants and safe milkweed species on which to lay eggs and feed larvae is imperative along the migration paths.

Monarch Migration Corridors

A small library could be filled with books, magazine articles, and science journals about the fall migrations of monarch butterflies. It's now treated as common knowledge that the eastern population (occurring east of the Rocky Mountains) migrates south, to the sanctuary of Michoacán forests in southwestern Mexico. There they try to survive the winter in a protected microclimate that, ideally, does not get too wet or too cold. In the West, most monarchs travel to the California coast in search of these same conditions in tall tree canopies near the ocean.

Ever enigmatic, however, monarchs may also be going other places. In recent years it's been noted that some autumn monarchs are traveling to areas in the Arizona desert (and perhaps other areas to the south within Mexico) in search of optimal overwintering conditions.

Less understood is the possibility that numbers of East Coast monarchs may be migrating to coastal overwintering sites in South Carolina or Georgia, or traveling farther south to Florida, Cuba, and the Caribbean islands. Only periodically, these inscrutable travels are observed and documented. In one especially mysterious case, in September of 1970, thousands of monarchs suddenly appeared in Bermuda, more than 600 ocean miles (1,000 km) away from the East Coast of the United States.

Autumn Rot & Parenthood

MILKWEEDS LOOK SICKLY IN FALL, as a general rule. Along with the other indignities they bear, the plants can be riddled with disease. They usually keep up appearances through summer; after flowering, however, seed making takes much of the energy a plant can muster.

At this point, it can be hard to look good. Hidden maladies become obvious. Cool weather and rainfall only make matters worse, especially when it comes to fungal and bacterial infections.

The most visible milkweed plague is leaf spot disease, caused by any of almost a dozen different fungal pathogens. Other fungi inscribe cankerlike lesions on stems and leaves. Still more types can produce general, whole-plant blight in which foliage suddenly turns yellow or brown, wilts, and ultimately withers into a dry, dead mass.

Milkweeds also contract several kinds of powdery mildew. As the name suggests, this manifests itself in thick coatings of white mildew on leaf surfaces. Much of that mildew mass is made up of dense spores that insects may pick up and spread among plants. Still more fungi live off the powdery mildew itself.

Other fungi rot the roots of milkweeds, causing them to soften into mush or fall off as brittle dead sections. Underground insects, feeding on the roots, probably exacerbate these diseases.

Rust Fungi: Deceptive and Diabolical

Of this rogue's gallery of milkweed diseases, rust fungi have the most complicated life cycle. They are often superficially visible as an orange or black coating on the undersides of leaves. Their delicate, flaky appearance can seem almost benign, revealing little about what the rust fungi are actually doing within the infected plant.

MILKWEED PLAGUES

FOLIAGE DISEASES AND THEIR SYMPTOMS

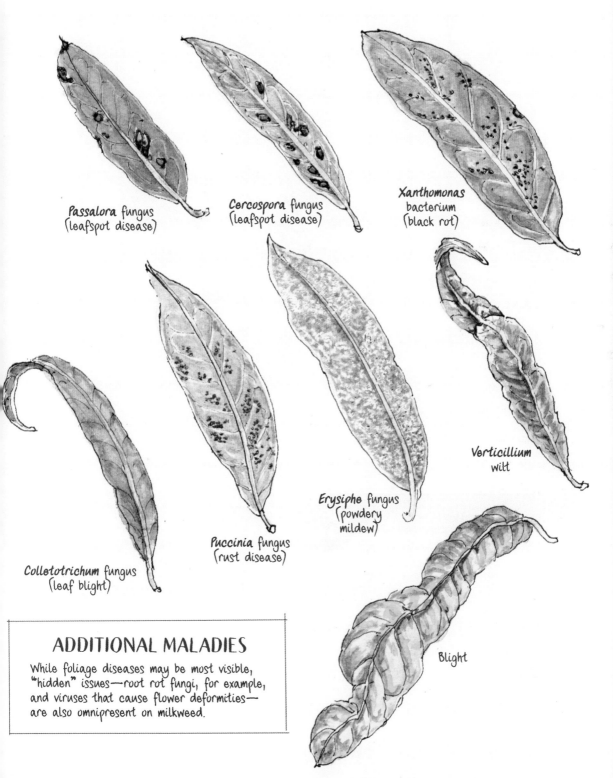

Passalora fungus
(leafspot disease)

Cercospora fungus
(leafspot disease)

Xanthomonas
bacterium
(black rot)

Colletotrichum fungus
(leaf blight)

Puccinia fungus
(rust disease)

Erysiphe fungus
(powdery
mildew)

Verticillium
wilt

Blight

ADDITIONAL MALADIES

While foliage diseases may be most visible,
"hidden" issues—root rot fungi, for example,
and viruses that cause flower deformities—
are also omnipresent on milkweed.

LIFE STAGES OF RUST FUNGI

During a typical year, rust produces different spore types, depending on the growth phase of its hosts.

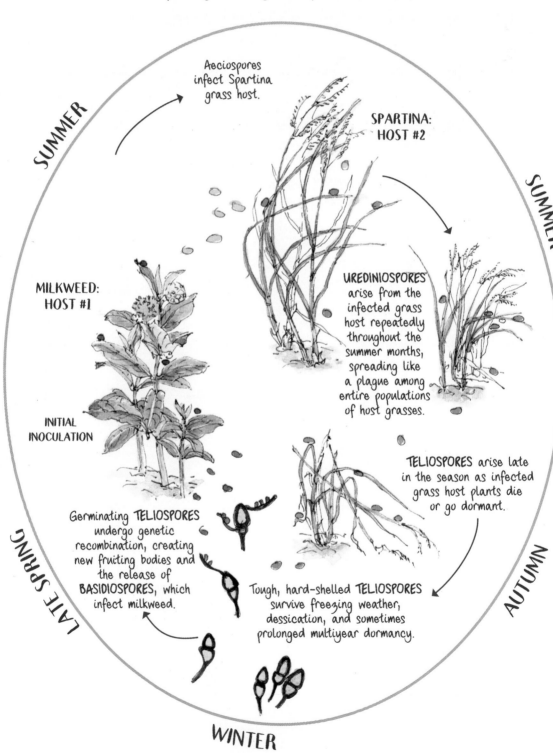

Aeciospores infect Spartina grass host.

SPARTINA: HOST #2

SUMMER

SUMMER

MILKWEED: HOST #1

UREDINIOSPORES arise from the infected grass host repeatedly throughout the summer months, spreading like a plague among entire populations of host grasses.

INITIAL INOCULATION

TELIOSPORES arise late in the season as infected grass host plants die or go dormant.

Germinating TELIOSPORES undergo genetic recombination, creating new fruiting bodies and the release of BASIDIOSPORES, which infect milkweed.

Tough, hard-shelled TELIOSPORES survive freezing weather, dessication, and sometimes prolonged multiyear dormancy.

LATE SPRING

AUTUMN

WINTER

Moreover, rusts often have two different host plants on which they live out their near-constant cycles of conquest and reproduction. It's likely that milkweeds and prairie grasses share the same rusts.

Rust spores ride on wind currents, infecting plants that are great distances, even oceans, apart. It's interesting that rust contagions among wild plants receive little attention in comparison to rusts of cultivated crops. In both cases, the infections can grow to epidemic proportions from just a single microscopic spore landing on the leaf surface of an appropriate plant species.

The odds of this happening in just the right way, and at the right time, must be one in a million. Yet it happens a lot.

A germinating rust spore produces a single tendril. This creeps along a leaf surface until it locates a pore (stoma), such as the kind used for plant respiration. From there it uses the pore as a migration corridor to the inner spaces between leaf cells.

Inside, rusts grow expanding rootlike structures known as haustoria, penetrating the tough cellulose walls of individual cells. This is a rather amazing feat, given that cellulose is composed of dense repeating units of glucose molecules arranged somewhat like the chemical equivalent of brickwork. Yet rusts secrete enzymes that dissolve the equivalent of mortar between those bricks.

· · · · ·

WITH THE OUTER WALLS BREACHED, haustoria are able to capture the incoming flow of water and nutrients destined for the plant's cells and to divert those nutrients for their own growth. With this food source, rusts can then expand their attack and begin spore production.

Space Flowers

Spore complexity is one of the things that make rust diseases so fascinating and devastating. Depending on environmental and host-plant cues, rusts can produce different spore types. Some of these may be ready to germinate immediately; others are designed to remain dormant through winter and drought. Rust spores may be specific to the same host plant on which they were born; still others are designed to infect an alternative host plant.

Since first seeing them in an ancient textbook, I've thought the spore-producing structures of rust fungi resemble what flowers would look like on another planet—both familiar and otherworldly.

When these "space flowers" bloom, a host plant may generate billions of spores. These spore clouds are scattered to the winds and rain. When whole fields of plants, especially grasses, are infected, airborne spore density might exceed 2.5×10^7 per square meter.

Untold numbers of them surely fall into the oceans. Others slowly desiccate in random, unrewarding places—maybe as dust behind your refrigerator. Others are fed on by woodlice and other small invertebrates.

We think little of these individual microscopic specks of dormant life, each possessing a kind of limitless ferocity to germinate and produce a trillion more of itself. Yet, periodically throughout human history, pandemic cycles of rusts in food crops have occurred and caused starvation and hardship. Wheat rust in particular has been our companion since agriculture began and has been implicated in the decline of Roman society—despite blood sacrifices of red-colored animals to Robigus, the god of rust.

INVASION BY RUST FUNGI

HOW A MILKWEED LEAF GETS INFECTED

Wind blows

1. A germinating spore produces a tendril that extends down the nearest stoma (plant pore) to enter the inner leaf.

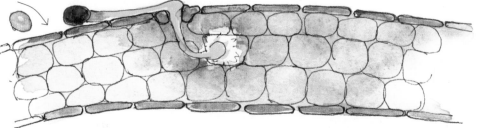

A CROSS SECTION OF A LEAF

After penetrating a cell wall, the rust can divert plant nutrients to sustain itself.

2. The infection spreads: The rust produces rootlike **HAUSTORIA** that penetrate deeper into the plant, extracting nutrients.

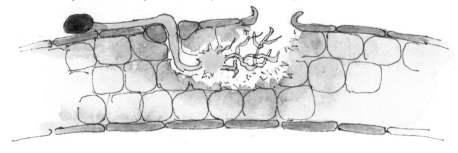

A CROSS SECTION OF A LEAF

3. Upon extracting enough energy for new growth, the rust produces new fruiting bodies on the leaf surface, and releases a new wave of windborne spores.

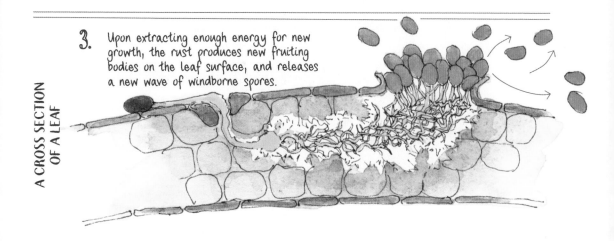

A CROSS SECTION OF A LEAF

OTHER PATHOGENS

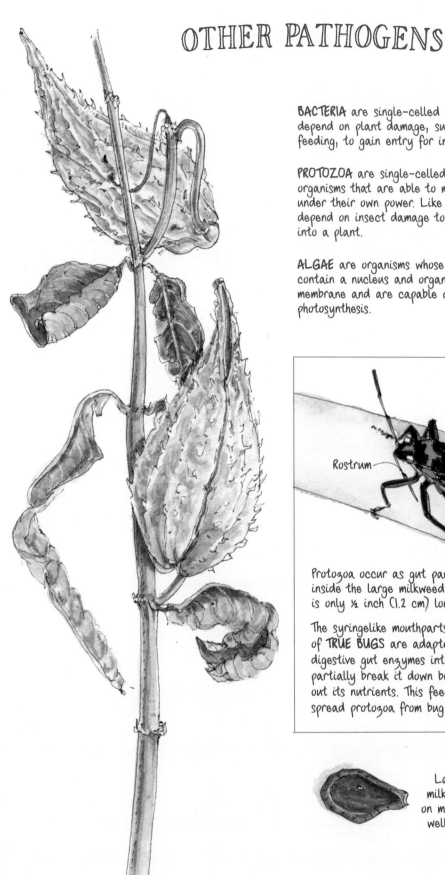

BACTERIA are single-celled organisms that depend on plant damage, such as insect feeding, to gain entry for infection.

PROTOZOA are single-celled animal-like organisms that are able to move around under their own power. Like bacteria, they depend on insect damage to gain entry into a plant.

ALGAE are organisms whose plantlike cells contain a nucleus and organelles within the membrane and are capable of carrying out photosynthesis.

Rostrum

Protozoa occur as gut parasites inside the large milkweed bug, which is only ½ inch (1.2 cm) long.

The syringelike mouthparts (**ROSTRUM**) of **TRUE BUGS** are adapted for injecting digestive gut enzymes into a plant, to partially break it down before suctioning out its nutrients. This feeding process can spread protozoa from bug to plant.

Large and small milkweed bugs feed on milkweed seeds as well as other plant tissue.

Plant Tribulations

Cool autumn rains also bring outbreaks of bacterial diseases, and these are less sophisticated than fungal contagions in how they mount an attack on plants. Unlike fungi, bacteria depend more on open wounds, created by feeding insects, for points of entry. These infections cause ugly, irregular dead spots on leaves. They also ooze out replicating bacterial cells (known as bacterial streaming), further spreading an infection through rainwater splashing into other wounds and pores.

At least one species of algae grows on milkweed leaves, causing even more kinds of leaf spots. There are protozoa—single-celled mobile organisms—that live alternately within the midgut of large milkweed bugs and the milkweed plants they feed on. They travel back and forth in the backwash between bug and plant, and from plant to plant, through the milkweed bugs' syringelike mouths.

Taken as a whole, milkweeds live in a near-constant state of exposure to highly contagious diseases, to ongoing cycles of plant pandemics. They are devoured inside and out at nearly every stage of their life. We tend to ignore the tribulations of plants, especially of ignoble ditch weeds. However, plants can and do become extinct from these travails, especially when such human abuses as mowing and herbicides are added to the mix. Instead, if we notice at all, we just see that the milkweed looks crummy in fall.

Seeds and Hope

It must be the plant kingdom's version of optimism: Even if milkweeds live with perpetual decay and sickness, they also make seed. Despite clear evidence that things never improve, milkweeds can be prolific seed producers when visited by enough pollinators.

Milkweed seeds develop within one of a pair of carpels (the inner chambers that enclose the ovules). Usually only one of this pair will develop; the other, if fertilized, may abort. As they grow, the seeds overlap in a cylindrical configuration that resembles fish scales. In early stages, the seeds are fleshy and green, and the follicle skin is leathery, flexible, and resistant to puncture.

In some species, young green follicles readily leak toxic latex from hundreds of points as a response to even a gentle touch. As a boy, I knew that touching the young pods would make my hands sticky, but I inevitably did it anyway. If I recalled this urge more often as I watch my own kids, I would be a better parent.

Depending on the species, follicles can be narrow and slender, or wide and spindle shaped. They can be smooth or bumpy or even covered with woolly hairs. They tend to grow at sharp angles away from the main stem. As the seeds within develop, the follicles can twist in orientation while the stem connection dries out and becomes dead, woody tissue.

Eventually, the follicle skin dries and splits lengthwise. As the brittle opening widens, revealing the flat brown seeds within, wind and rain begin to stir the exposed coma—the silky white filament hairs attached at the seeds' pointed ends. With the coma of each seed tucked under another, the effect of the wind is similar to that of a zipper pulling to release the entire seed cluster into the open air.

INSIDE A MILKWEED POD

The **SEED** is an embryonic plant, formed from a ripened ovule.

Up to around 150 seeds may be found within a follicle.

A **FRUIT** is the seed-bearing structure that forms from a plant's ovaries.

The **FOLLICLE** or pod is a dry, one-chambered fruit that opens along the opposite side from where the stem is attached.

Asclepias syriaca (common milkweed [very spiny])

SOME FOLLICLE SHAPES

Asclepias latifolia (broadleaf milkweed)

Asclepias speciosa (showy milkweed)

Asclepias exaltata (poke milkweed)

Asclepias asperula (antelope-horns milkweed)

Asclepias fascicularis (narrowleaf milkweed)

SEED DISPERSAL PATHS

WIND
Moves seeds unknown
distances

ANIMALS
Move seed for many
feet or many miles

WATER
Moves seed for many
feet or many miles

How Seeds Travel

Folk wisdom and kids' botany books tell us that milkweeds are wind dispersed—lifted and carried by the sail effect of the large fluffy coma, floating great distances over land. Yet there are reasons to scrutinize this theory, including that coma hairs are surprisingly brittle. The fibers quickly break, most often where the coma attaches to the seed.

There also seems to be a trade-off between dispersal distances and seed viability. In a study conducted in the 1980s, researchers found that among seeds with air-catching structures, the smaller ones floated farther on wind currents but had lower germination and survival rates. Also, all else being equal, taller plants disperse seeds farther on wind currents than short plants do. Nevertheless, a lot of milkweed species are short statured and hidden among taller vegetation.

It turns out that seeds we assume to be wind dispersed often get more mileage from being stuck to animal fur. Having watched these plants most of my life, I wonder whether most windborne milkweed seeds travel only a few dozen feet from their mother plant.

Whether or not milkweeds are long-distance air travelers, some are surprisingly adept at journeying by water. Decades ago, weed scientists in eastern Washington State noticed that showy milkweed often appears along irrigation ditches, where matted seeds and coma would drift up and down the lengths of the ditches, through culverts and drainpipes. Testing revealed that showy milkweed seeds can be submerged in water for months and still quickly germinate and grow, often more rapidly than seeds stored in dry conditions.

Aquatic milkweed takes this strategy further. Like swamp milkweed, *Asclepias perennis* is a muck plant, but one native to the lower Mississippi River valley, where it occupies cypress swamps and wet ditches. Rare among milkweeds, it produces no coma at all. Instead, it drops bare, buoyant seeds that float on water, sometimes for months, moving on river currents and floods. To facilitate this, aquatic milkweed grows follicles that droop downward as they develop, to drop the seeds more easily and directly into the water around the mother plant.

Seed Shatter and Seed Rain

Seed shatter, as it is called when follicles burst and release their seeds to the winds and waters, is part of a larger autumn phenomenon known as seed rain. It's the collective effect of massive numbers of plants of all kinds—grasses and forbs—releasing their seeds together. The weight of this can be hundreds or thousands of pounds of seed dropped per acre in a prairie or abandoned weedy brownfield.

Much of this seed mass accumulates in the thatch layer, often lacking adequate soil contact for germination. If an opportunity to germinate doesn't arrive, eventually these seeds are covered by years of accumulated dead vegetation, becoming buried far underground, sometimes still viable but too cold and light deprived to germinate. In some places, deep beneath lawns and roads, there are ancient seed banks of this kind, some hundreds of years old, waiting to be excavated and re-create ancient ecosystems that we can barely imagine.

SEED RAIN

Ancient Fires

Fall is also the time of fire, at least in the places I've lived. Rank stands of thick, dead autumn grass go up as tinder. Scientists in eastern Europe, trying to understand the invasive spread of common milkweed across their continent, found that exposure to wildfire smoke increased the germination of seeds. Perhaps this makes sense, given that the plant has coexisted with burned earth for a long time.

In the Midwest, the ancient tradition of burning prairies, once the practice of Native people, has experienced a small-scale revival by prairie ecologists. Cleared of dense thatch, burned prairies allow light to reach the seedbed and create more space for wildflowers otherwise smothered by grasses.

Farther west, we live with a deeper fear of fire. Fueled by drought and dying forests, fire season keeps us in our houses, windows taped shut against choking outdoor air and unnerving orange skies, curtains drawn so the kids don't see what's out there. Whatever value smoke might have for seed germination is counteracted by an absence of rain.

Still, the milkweeds that remain try to make seed.

When I was not much older than my own sons are now, still single digits of age, I often played in an old dumping ground used by the town road crew. It was a place of box elders, thistles, and common milkweed amid piles of heaped concrete slabs, jagged in shape and halfway embedded in the ground.

This was the 1980s, and I was a latchkey kid with a Kool-Aid mustache. The slabs were my imaginary bomb shelters, a play space predicated on nuclear war, which I knew nothing about except from overheard conversations between adults. As a parent, I now wonder what adult exchanges my own sons overhear.

PRAIRIE FIRES

RENEWAL

THE ACCIDENTAL GARDEN

1. *Asclepias syriaca* (common milkweed)
2. *Prunus virginiana* (chokecherry)
3. *Ulmus pumila* (Siberian elm)
4. *Elymus repens* (quackgrass)
5. *Solidago canadensis* (Canada goldenrod)

MY OWN GROWN-UP THOUGHTS NOW LINGER over the future of untamed plants and animals, those that still resist humanity's demand that they bow down to us. How much of that giddy and ungovernable garden will there be in my sons' future world? Humans are a transcendent, grand, and marvelous species. We also treat other living things as something to own, sell, or act against, according to our whims. I'm no exception.

An acquaintance tells me that people have been forecasting doom and gloom since the beginning of time. Perhaps he's right. Maybe the deeper mystery is that we all—human and milkweed alike—just keep going as long as we can, even as the unnatural becomes natural.

The Joy of Meadows

Most land disasters start over as a meadow. Meadows are usually the first plant community to spring up after a forest fire or a bomb blast. Described as "early successional habitat" by ecologists, they are the same plant communities that grow up around fallen factories and razed city blocks with contaminated soils.

Also known as grasslands, old fields, prairies (from the French word for meadow), and more obscure terms, meadows are typically dominated by three plant families: Poaceae (grass), Asteraceae (sunflower), and Fabaceae (pea or bean), with assorted others, like milkweed, scattered throughout. Of these, the grasses are generally the most abundant, although in a healthy meadow all the plants contribute something to one another.

We often overlook meadows when considering nature and conservation. Forests, with their darkness and drama, loom in the imagination and mobilize worthy sympathies for their protection. The scale of mountains and oceans requires us to recognize them as wilderness beyond our full grip. But meadows are an afterthought.

This is okay, because meadows don't mind if we overlook them. Despite us, they are brilliantly full of living things, from microbes to mammals to plants that spread their seeds in amazing ways. They support this life while also functioning as humanity's janitorial service—capturing our waste, whether factory effluent, cigarette butts, or old sofas—and turn it all into something green.

Meadows sequester enormous amounts of carbon. They give us somewhere to walk. They allow us to spot enemies from far away. They give us ample spaces to try out dubious ideas.

Meadows and humans thrive together. We may doubt this. It's true that they suffer constantly from invasive species, altered drainage, pollution and dumping, urban development, fragmentation, periodic fires, grazing (and other agricultural attempts), and the boorish human instinct to occupy any grassy piece of land. We have built great Midwest cities on meadows. We have built racetracks over them and big-box stores and nuclear waste dumps.

But meadows always come back. Every crack in the pavement sprouts a new one: at first only a single blade of grass, then other blades, in other cracks. Eventually the cracks converge, and the whole riotous affair starts to swallow up our rubbish, paying us back with grasshoppers and flowers.

MEADOWS

...are open lands of grasses,
herbaceous perennials and annuals, and other nonwoody plants.

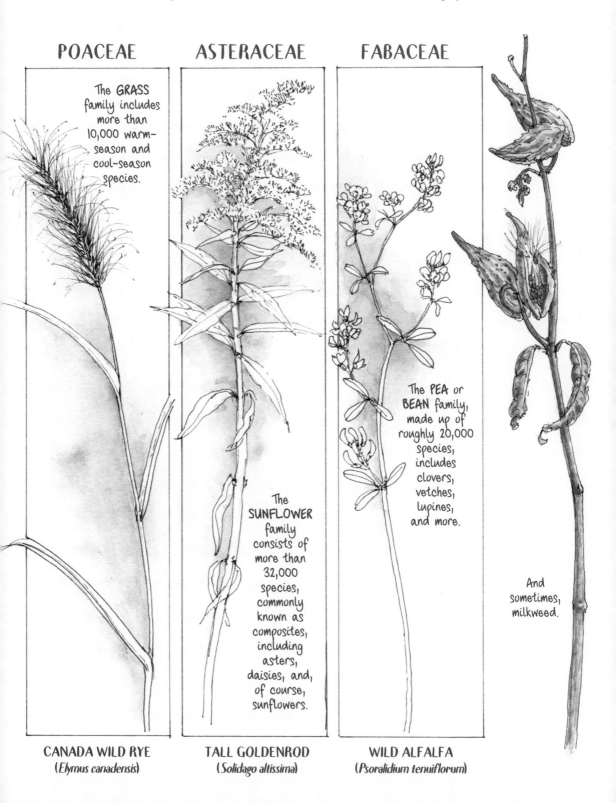

POACEAE

The GRASS family includes more than 10,000 warm-season and cool-season species.

ASTERACEAE

The SUNFLOWER family consists of more than 32,000 species, commonly known as composites, including asters, daisies, and, of course, sunflowers.

FABACEAE

The PEA or BEAN family, made up of roughly 20,000 species, includes clovers, vetches, lupines, and more.

And sometimes, milkweed.

CANADA WILD RYE
(*Elymus canadensis*)

TALL GOLDENROD
(*Solidago altissima*)

WILD ALFALFA
(*Psoralidium tenuiflorum*)

Meadows are our ecosystem. It's where our own species originated, at least according to the ecological theory known as the Savanna Hypothesis. Within them we invented agriculture and learned to master fire. We built the starter homes of civilization from prairie sod and the skins of wild grassland beasts. We fashioned fibers out of long pliable grasses and other meadow plants. Almost the entirety of human existence has had us living in or around meadows.

Take comfort in this. That a living system can bear what meadows do is nothing short of amazing. Responding to abuse or disaster, Earth's meadows just morph into new versions of themselves, sometimes in new places. This is the most remarkable natural history story that I know.

Meadows don't go away easily. They adjust things a little, adapt, and accidentally look beautiful in the process—like the milkweeds within them.

Like all of us.

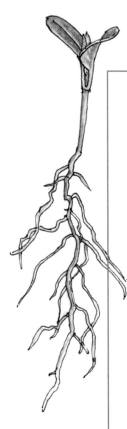

HOW TO START MILKWEED FROM SEED

Given the diversity of species and the variability within species, there is no single perfect formula for how to grow your own milkweed from seed. Nevertheless, here are some general strategies that can help maximize success.

Many milkweeds, especially from northern locations, benefit from cold-moist stratification. The precise length of cold exposure needed to break dormancy and promote germination is usually impossible to gauge, especially with milkweed populations that you have not already worked with.

In most cases, a good compromise is to mix the seeds with a small amount of very slightly damp sand, perlite, peat moss, or vermiculite. Place the resulting mixture in a sealed bag in your kitchen refrigerator for four to six weeks.

Upon removing the seeds from your refrigerator, inspect them for signs of germination. Any that have started germinating can be immediately transplanted.

For the rest, rinse them to wash off the sand or other stratification material, and then soak them in hot (but not boiling) water. Hot tap water is usually perfect.

Let the water cool to a comfortably warm temperature—a temperature that you would give a child or a pet a bath in. Try to maintain that approximate temperature for two to three days. A heating pad underneath your soaking bowl can help.

Change the water daily, and watch for signs of germination. Immediately transplant any seeds that start germinating into warm, fertile, loose soil.

After three days, loosely wrap any seeds that still have not germinated in slightly wet paper towels, place them inside a plastic bag, and keep the bag on a warm countertop. Check them daily for germination, and transplant them at the first sign of an emerging root.

Index

References to illustrations are in *italics*.

A

ABC and XYZ of Bee Culture, The, 14
accidental garden, *106*
Ag PhD, 12
algae, *96*, 97
American Apiculturist, 14
Anderson, John, 41, 42, 44, 47, 59
ants, and aphids, *55*, 56
aphids, 48–56
aquatic milkweed, 102
Asclepias genus, 2, 3
asexual reproduction, *49*

B

bacteria, *96*, 97
bee-pastures, 38, *40*
beetle plague, 44–47
black-headed grosbeak, 66
bumble bees, *28*, 29, 83–84
butterfly milkweed, 14

C

cache, *23*
cardenolides, 7, 8
Carver, George Washington, *10*
cell structure, *32*
Central Valley (California), 38–43, *58*, 59
clones, *51*
cobalt milkweed beetle, 44–47
coloration, 8, *9*
common milkweed, 11, 14
Cornut, Jacques-Philippe, 2
crown flower, 4

D

diseases, 90–97
ditch weeds, 62, *63*, 66–69
dodder, 30
dogbane family, 3–6

E

eating milkweed, 8, *10*, 11
  See also milkweed feeders
eradication of milkweed, 12–14

F

fish, *76*, 77
floss, 15–17, 20, 22
follicles, 81, 98
  See also pods

G

germination
  starting milkweed from seed, 112
  and temperature, 33–34
  and wildfire, 104
Great Plains, Northern, *35*

H

Hébert, Louis, 2
Heddon, James, 14
Hedgerow Farms, 42, 44, 47
hedgerows, 41, *43*, *58*, 59
herbicides, 17
honey production, 14

L

leaf spot disease, 90
Linnaeus, Carl, 2–3

M

meadows, 107–110
mice, 20–22
Midwest, Upper, 56–57
migration, 72–77, *86*, 87
milkweed bugs, 26, *27*
milkweed feeders, *64*, 65–66, *67*
  See also eating milkweed
Milkweed Floss Corporation of America, 15
milkweed flowers, 78–84
  morphology, *79*
milkweed longhorn beetles, 30
milkweeds, representative species, *6*
Mississippi River, Upper, 70–72
  fish of, *76*
mole crickets, 25
moles, 25